OFF-HAND SKETCHES.

A little dashed with Humour.

By T. S. ARTHUR.

NEW YORK
JOHN W. LOVELL COMPANY
14 AND 16 VESEY STREET

TROW'S
PRINTING AND BOOKBINDING COMPANY,
NEW YORK.

PREFACE.

The reader cannot but smile at some of the phases of life presented in this volume. Yet the smile will, in no case, the author thinks, be at the expense of humanity, good feeling, or virtue. Many of the incidents given, are facts embellished by a few touches of fancy. In all, lessons may be read that some, at least, will do well to lay to heart

CONTENTS.

OFF-HAND SKETCHES.

THE CIRCUIT-PREACHER.

THE Methodist circuit-preacher is in the way of seeing human nature in many rare and curious aspects. Under the itinerating system, the United States are divided into conferences, districts, and circuits. The conference usually embraces a State, the district a certain division of the State or conference, and the circuit a portion of the district. To every circuit is assigned a preacher, who is expected to provide himself with a horse, and his duty is to pass round his circuit regularly at appointed seasons through the year, and meet the members of the church at the various places of worship established on the circuit. Every year, he attends the annual conference of preachers, at which one of the bishops presides, and is liable to be assigned a new circuit, in the selection of which, as a general thing, he has no choice—the bishop making all the appointments;

and so, term after term, he goes to a new place, among strangers. Before any strong attachments can be formed, the relation between him and his people is severed; and he begins, as it were, life anew, hundreds of miles away, it may be, from any former field of labour. To a married man, this system is one involving great self-denial and sacrifice, assuming often a painful character.

In those circuits that embrace wealthy and populous sections of the country, the Methodist minister is well taken care of; but there are many other sections, where the people are not only very poor, but indifferent to matters of religion, ignorant in the extreme, and not over-burdened with kind or generous feelings. On circuits of this character, the preacher meets sometimes with pretty rough treatment; and if, for his year's service, he is able to get, being, we will suppose, a single man, fifty or sixty dollars in money, he may think himself pretty well off.

To one of these hard circuits, a preacher, whom we shall call the Rev. Mr. Odell, of the New Jersey conference, found himself assigned by the bishop who presided at the annual conference. The change was felt as pretty severe, he having been on a comfortable station for two years; but as he must take the evil with the good, he conscientiously repressed all natural regrets and murmurings, and, as in duty bound, started, at the close of the conference, for his new field of labour. A day or two before leaving, and after the appointments were announced, Mr. Odell said to the brother who had ridden that circuit during the previous year—" So, I am to follow in your footsteps?"

" It appears so," was the brief reply.

" How did you like the circuit?"

" I am very well pleased to change."

" Not much encouragement in that answer."

" We can't all have good places. Some of us must take our turn in the highways and byways of the land."

" True; I am not disposed to complain. I have taken up the cross, and mean to bear it to the end, if possible, without a murmur."

"As we all should. Well, brother Odell, if you pass the year on the circuit without a murmur, your faith and firmness will be strong. I can assure you that it will be more than I did—a great deal more."

" I have been among some pretty rough people in my time."

" So have I; but"— and he checked himself; " however, I will not prejudice your mind; it would be wrong. They do as well, I suppose, as they know how, and the best can do no more."

" Truly said. And the more rude, ignorant, and selfish they are, the more need they have of gospel instruction, and the more willing should we be to break the bread of life for them. If our Master had not even 'where to lay his head,' it ill becomes us to murmur because every natural good is not spread out before us."

In this state of mind, Odell went to his new circuit. Having deposited his family, consisting of a wife and one child, in the little village of S——, with a kind brother, who offered them a home at a mere nominal board, he mounted his horse and started forth on a three weeks' tour among the members of the church to whom he was to minister, during the next twelve months, in holy things. The first preaching-place was ten miles distant, and the little meeting-house stood on the roadside, nearly a mile from any dwelling, and in an exceedingly poor district of country.

Before leaving S——, Mr. Odell made inquiries of the brother at whose house he was staying, in regard to the route he was to take, and the people among whom he was going. As to the route, all that was made satisfactory enough; but the account given of the people was not encouraging in a very high degree.

"The fact is," said the brother, rather warmly, "it's my opinion that they don't deserve to have the gospel preached among them."

To this, however, the preacher very naturally demurred, and said that he was not sent to call the righteous, but sinners, to repentance.

"Where will I stop to-night?" he inquired. It

was Saturday afternoon, and on Sunday morning he was to preach at his first appointment.

"Well," said the brother, slowly and thoughtfully, "I can tell you where you ought to stop, but I don't know you will be so welcome there as at a poorer place. Brother Martin is better able to entertain the preachers comfortably than any one else in that section; but I believe he has never invited them home, and they have generally gone to the house of a good widow-lady, named Russell, whose barrel of meal and cruse of oil deserve never to fail. She is about the only real Christian among them."

"Is brother Martin a farmer?"

"Yes, and comfortably off; but how he ever expects to get his load of selfishness into heaven, is more than I can tell."

"You must not be uncharitable, brother," said Odell.

"I know that; but truth is truth. However, you must see and judge for yourself. I think you had better go to the house of sister Russell, who will welcome you with all her heart, and give you the best she has."

"And I want no more," said the preacher.

After getting precise directions for finding sister Russell, he started on his journey. It was nearly five o'clock, and he made his calculation to reach sister Russell's by seven, where he would remain all night, and go with her to the preaching-place on Sunday morning. He had not, however, been half an hour on his journey, before heavy masses of deep blue clouds began to roll up from the horizon and spread over the sky; and ere he had accomplished half the distance he was going, large drops of rain began to fall, as the beginning of a heavy storm. The preacher was constrained to turn aside and seek the shelter of a farm-house, where he was received with much kindness.

Night-fall brought no abatement of the tempest. The lightning still blazed out in broad masses of fire, the thunder jarred and rattled amid the clouds like parks of artillery, and the rain continued to pour down unceasingly. The invitation to remain all night, which the farmer and his wife tendered in all sincerity, was not, of course, declined by the preacher.

In the morning, after being served with a plentiful breakfast, Odell returned his warmest thanks for the kindness he had received, and proceeded on his journey. He had five miles to ride; but it was only half-past eight o'clock when he started, and as the hour for preaching was ten, there was plenty of time for him to proceed at his leisure. As sister Russell lived nearly a mile away from a direct course, he did not turn aside to call upon her, but went on to the meeting-house. On reaching the little country church, Mr. Odell found a small com-party of men assembled in front of the humble building, who looked at him curiously, and with something of shyness in their manner, as he rode up and dismounted. No one offering to take his horse, he led him aside to a little grove and tied the reins to a tree. One or two of the men nodded, distantly, as he passed them on his way to the meeting-house door, but none of them spoke to him.

On entering the meeting-house, Mr. Odell found some thirty persons assembled, most of them women. If there were any "official members" present, they made themselves in no way officious in regard to the preacher, who, after pausing at the door leading into the little altar or chancel for a short time, and looking around with an expression of inquiry on his face, ascended the pulpit-stairs and took his seat. All was as silent, almost, as if the house had been tenantless.

In a little while, the preacher arose and gave out a hymn; but there was no one to raise the tune. One looked at another uneasily; sundry persons coughed and cleared their throats, but all remained silent. Odell was not much of a singer, but had practised on "Old Hundred" so much, that he could lead that air very well; and the hymn happening, by good luck, to be set to a long-metre tune, he was able to start it. This done, the congregation joined in, and the singing went off pretty well. After praying and reading a chapter in the Bible,

Odell sat down to collect his thoughts for the sermon, which was, of course, to be extempore, as Methodist sermons usually are. It is customary for the choir, if there is one, to sing an anthem during this pause; or, where no singers are set apart, for some members to strike up an appropriate hymn, in which the congregation joins. On this occasion, all was silent. After the lapse of a few minutes, Mr. Odell arose, and turning, in the Bible, to the chapter where the text, from which he was to preach, was recorded, read the verse that was to form the groundwork of his remarks. Before opening the subject, he stated, briefly, that he was the preacher who was to labour among them during the ensuing year, and hoped, in the Divine Providence, that good, both to them and to him, would result from the new spiritual relations that were about to be commenced. Then proceeding with his discourse, he preached to and exhorted them with great earnestness, but without seeming to make any impression. Not an "amen" was heard from any part of the house; not an eye grew moist; not an audible groan or sigh disturbed the air. Nothing responded to his appeals but the echo of his own voice.

Never had the preacher delivered a discourse in which he felt so little freedom. His words came back upon his ears with a kind of a dull reverberation, as if the hearts of his hearers were of ice, instead of flesh.

Before singing the last hymn, which Mr. Odell gave out at the conclusion of the sermon, he announced that he would hold a class-meeting. After he had finally pronounced the benediction, there was a general movement towards the door; only seven remained, and these were all female members, most of them pretty well advanced in their life-journey. Mr. Martin was at the meeting, but ere the preacher had descended the pulpit-stairs, he was out of the house and preparing to leave for home.

"Where is the new preacher going?" asked a member, of Mr. Martin, as he led out his horse.

"To sister Russell's, I presume."

"Sister Russell is not here."

"Isn't she?"

"No; she's sick."

"He stayed there last night, I suppose, and will go back after class." Martin sprang upon his horse as he said this.

"We ought to be sure of it," remarked the other.

"I can't invite him home," said Martin. "If I do, I shall have him through the whole year, and that is not convenient. The preachers have always stayed at sister Russell's, and there is no reason why they shouldn't continue to do so."

"I haven't a corner to put him in," remarked the other. "Besides, these preachers are too nice for me."

"It's all right, no doubt," said Martin, as he

balanced himself in his saddle; "all right. He stayed at sister Russell's last evening, and will go back and stay there until to-morrow morning. Get up, Tom!" And, with this self-satisfying remark, the farmer rode away.

The man with whom he had been talking, was, like him, a member; and, like him, had omitted to attend class, in order to shift off upon some one else the burden of entertaining the new preacher; for whoever first tendered him the hospitalities of his house and table would most probably have to do it through the year. He, too, rode off, and left others to see that the preacher was duly cared for. An icy coldness pervaded the class-meeting.

Only four, out of the seven sisters, one of whom was an old black woman, could muster up courage enough to tell, in answer to the preacher's call, the "dealing of God" with their souls; and only two of them could effect an utterance louder than a whisper. What they did say had in it but little coherence, and Mr. Odell had to content himself with an exhortation to each, of a general rather than a particular character. When the hymn was sung at the close, only one thin voice joined in the song of praise, and not a sob or sigh was heard in response to his prayer. The class-paper showed the names of thirty members, but here were only seven! This was rather discouraging for a commencement. Mr. Odell hardly knew what course to take; whether to stir up with some pretty sharp remarks the little company of believers who were present, and thus seek to impress the whole through them; or to wait until he came round again, and have a good chance at them from the pulpit. He concluded in the end, that the last course might be the best one.

In calling over the names on the class-paper he found that sister Russell was absent. On dismissing the meeting, all except the old black woman retired. She lingered, however, to shake hands with the new preacher, and to show him that, if she was old, her teeth were good, and her eyes bright and lively.

On emerging into the open air, Odell saw the last of his flock slowly retiring from the scene of worship. For two of the women, their husbands had waited on the outside of the meeting-house, and they had taken into their wagons two other women who lived near them. These wagons were already in motion, when the preacher came out followed by the old black woman, who it now appeared, had the key of the meeting-house door, which she locked.

"Then you are the sexton, Aunty," remarked Odell, with a smile.

"Yes, massa, I keeps de key."

"Well, Nancy," said Odell, who had already made up his mind what he would do, "I am going home to dinner with you."

"Me, massa!" Old Nancy looked as much surprised as a startled hare.

"Yes. You see they've all gone and left me, and I feel hungry. You'll give me some of your dinner?"

"Yes, massa, please God! I'll give you all of it—but, it's only pork and hominy."

"Very good; and it will be all the sweeter becauso I am welcome."

"'Deed massa, and you is welcome, five hundred times over! But it was a downright shame for all de white folks to go off so. I never seed such people."

"Never mind, Nancy, don't trouble yourself; I shall be well enough taken care of. I'll trust to you for that."

And so Mr. Odell mounted his horse, and accompanied the old woman home. She lived rather over a mile from the meeting-house—and the way was past the comfortable residence of Mr. Martin. The latter did not feel altogether satisfied with himself as he rode home. He was not certain that the preacher had stayed at sister Russell's the night before. He might have ridden over from S—— since morning. This suggestion caused him to feel rather more uneasy in mind; for, if this were the case, it was doubtful whether, after class was over, there would be any one to invite him home.

"What kind of a man is the new preacher?" asked Mrs. Martin of her husband, on his return from meeting.

"He seemed like a very good sort of man," replied Martin, indifferently.

"Is he young or old?"

"He's about my age, I should think."

"Married?"

"I'm sure I don't know."

"Did you speak to him?"

"No, I came away after the sermon."

"Then you didn't stop to class?"

"No."

"Sister Russell was not there, of course?"

"No; she's sick."

"So I heard. The preacher didn't stay at her house last night."

"How do you know?"

"Mrs. Williams called in while you were away. She had just been to sister Russell's."

"And the new preacher didn't stay at her house last night?"

"No. Mrs. Williams asked particularly."

"He must have ridden over from S—— this morning. I am sorry I didn't wait and ask him to come home and stay with us."

"I wish you had. Sister Russell is too sick to have him at her house, if he should go there. Who stayed to class-meeting?"

"Not over half a dozen, and they were all women I left Bill Taylor and Harry Chester waiting outside for their wives."

"They wouldn't ask him home."

"No; and if they did, I should be sorry to have him go there. I wish I had stayed in, and invited him home. But it can't be helped now, and there's no use in fretting over it."

Soon after this, dinner was announced, and the farmer sat down with his family to a table loaded with good and substantial things. He ate and enjoyed himself; though not as highly as he would have done, had not thoughts of the new preacher intruded themselves.

After dinner, Martin took a comfortable nap, which lasted about an hour. He then went out and took a little walk to himself. While standing at the gate, which opened from his farm on to the county road, a man, who lived half a mile below, came along. This man was not a member of any church, and took some delight, at times, in having his jest with professors of religion.

"Fine afternoon, Mr. Ellis," said Martin, as the man stopped.

"Very fine. How are you all?"

"Quite well. Any news stirring?"

"Why, no, not much. Only they say that the Methodists about here have all joined the Amalgamation Society."

"Who says so?" inquired Martin, slightly colouring.

"Well, they say it down our way. I thought it

was only a joke, at first. But a little while after dinner, Aunt Nancy's Tom came over to my house for some oats and hay for your new minister's horse. He said the preachers were going to stop at the old woman's after this. I half-doubted the rascal's story, though I let him have the provender. Sure enough, as I came along just now, who should I see but the preacher sitting before the door of old Nancy's log-hut, as much at home as if his skin were the colour of ebony. These are rather queer doings, friend Martin; I don't know what folks 'll say."

We will not pause to describe the astonishment and confusion of Martin, on learning this, but step down to Aunt Nancy's, where Odell, after dining on pork and hominy, with the addition of potatoes and corn-bread, was sitting in the shade before the log cabin of the old negro. The latter was busy as a bee inside in preparation of something for the preacher's supper, that she thought would be more suited to his mode of living and appetite, than pork, corn-bread, and hominy.

Odell was rather more inclined to feel amused than annoyed at his new position. Aunt Nancy's dinner had tasted very good; and had been sweetened rather than spoiled by the old creature's loquacious kindness and officious concern, lest what she had to set before him would not be relished. While he thus sat musing—the subject of his thoughts is of no particular consequence to be known—his attention was arrested by hearing Aunt Nancy exclaim—

"Ki! Here comes Massa Martin!"

The preacher turned his head and saw a man approaching with the decided and rather quick step of one who had something on his mind.

"Is that brother Martin?" asked Mr. Odell, calling to Aunt Nancy, who was near the window of her hut.

"Yes, please goodness! Wonder what he comin' here 'bout."

"We'll soon see," returned the preacher, composing himself in his chair.

In a few minutes, the farmer, looking sadly "flustered," arrived at the door of the old negro's humble abode. Odell kept his seat with an air of entire self-possession and unconcern, and looked at the new comer as he would have done at any other stranger.

"Mr. Odell, the new preacher on this circuit?" said Martin, in a respectful manner, as he advanced towards the minister.

"Yes, sir," replied Odell, without rising or evincing any surprise at the question.

"I am very sorry indeed, sir! very sorry," began Martin in a deprecating and troubled voice, "that you should have been so badly neglected as you were to-day. I had no idea—I never once thought—the preachers have always stayed at sister Russell's—I took it for granted that you were there To think you should not have been invited home by any one! I am mortified to death."

"Oh, no," returned the preacher, smiling; "it is not quite so bad as that. Our good old sister here very kindly tendered me the hospitalities of her humble home, which I accepted gratefully. No one could be kinder to me than she has been—no one could have given me a warmer welcome."

"But—but," stammered forth Martin, "this is no place for a preacher to stay."

"A far better place than my Lord and Master had. *The foxes have holes and the birds of the air have nests, but the Son of man hath not where to lay his head.* The servant must not seek to be greater than his Lord."

"But my dear sir! my house is a far more suitable and congenial home for you," urged the distressed brother Martin. "You must go home with me at once. My wife is terribly hurt about the matter. She would have come over for you herself, but she is not very well to-day."

"Tell the good sister," replied Odell, affecting not to know the individual before him, "that I am so comfortable here, that I cannot think of changing my quarters. Besides, after Aunt Nancy has been so kind as to invite me home, and provide for both me and my horse, when no one else took the least notice of me, nor seemed to care whether I got the shelter of a roof or a mouthful of food, it would not be right for me to turn away from her because a more comfortable place is offered."

It was in vain that Martin argued and persuaded. The preacher's mind was made up to stay where he was. And he did stay with Aunt Nancy until the next morning, when, after praying with the old lady and giving her his blessing, he started on his journey.

When, at the end of four weeks, Mr. Odell again appeared at the little meeting-house, you may be sure he was received with marked attention. Martin was the most forward of all, and, after preaching and class-meeting—there was a pretty full attendance at both—took the minister home with him. Ever since that time, the preachers have been entertained at his house.

THE PROTEST.

Reader! did you ever have a visit from that dreaded functionary—that rod in pickle, held in terrorem over the heads of the whole note-paying fraternity, ycleped a notary? I do not mean to insult you: so don't look so dark and dignified. I am serious. If no—why no, and there let the matter rest, as far as you are concerned; if yes, why yes, and so I have an auditor who can understand me.

As for me, I have been protested. I say it neither with shame nor pride. Yes, I have suffered notarial visitation, and am still alive to tell the tale.

I was in business when the exciting event occurred, and I am still in business, and I believe as well off as I was then. But let me relate the circumstance.

When I first started in the world for myself, I had a few thousand dollars. In a little while, I found myself solicited on all sides to make bills. I could have bought fifty thousand dollars' worth of goods as easily as to the amount of five thousand dollars; and the smallest sum I have named was about the extent of my real capital. There was one firm importunate above the rest, and they were successful in getting me into their debt more heavily than I was to any other house. If I happened to be passing their store, I would be called in, with—

"Here, Jones, I want to show you something. New goods just in; the very thing for your sales."

Or—

"Ah! how are you, Jones? Can't we sell you a bill, to-day?"

They were for ever importuning me to buy, and often tempted me to make purchases of goods that I really did not want. I was young and green then, and did not know any thing about shelves full of odds and ends, and piece upon piece of unsaleable goods, all of which had to be paid for.

For two or three years, I managed to keep along, though not so pleasantly as if I had used my credit with less freedom. By that time, however, the wheels of my business machinery were sadly clogged. From a salesman behind my counter, I became a "financier." (!)

During the best hours of the day, and when I was most wanted in the store, I was on the street, hunting for money. It was borrow, borrow, borrow, and pay, pay, pay. My thoughts were not directed toward the best means of making my business profitable, but were upon the ways and means of paying my notes, that were falling due with alarming rapidity. I was nearly all the time in the delectable state of mind of the individual who, on running against a sailor, was threatened with being knocked "into the middle of next week." "Do it, for heaven's sake!" he replied—"I would give the world to be there."

On Monday morning, I could see my way through the week no clearer than this note-haunted sufferer. In fact, I lived a day at a time. On the first of each month, when I looked over my bill-book, and then calculated my resources, I was appalled. I saw nothing ahead but ruin. Still I floundered on, getting myself deeper and deeper in the mire, and rendering my final extrication more and more difficult.

At last, I found that my principal creditors, who had sold me so freely from the first, and to whom nearly the half of what I owed was due, began to be less anxious about selling me goods. They did not call me in, as of old, when I passed, nor did they urge me to buy when I went to their store. Still they sent home what I ordered; but their prices, which before were the lowest in the trade, were now above the average rates. I noticed, felt, and thought I understood all this. I had been careful not to borrow money from that firm; still, I was borrowing, somewhere, every day, and they, of course, knew it, and began to be a little doubtful of my stability.

At last, I was cornered on a note of a thousand

dollars, due this house. Besides this note, I had fifteen hundred dollars of borrowed money to pay. At nine o'clock, I started forth, leaving good customers in the store, to whom no one could attend as well as myself. By twelve o'clock, I was able to return my borrowed money, and had the promise of a thousand dollars by half-past one. Until half-past one I waited, when a note came from the friend who had promised the loan, informing me with many expressions of regret, that he had been disappointed, and, therefore, could not accommodate me.

Here was a dilemma, indeed. Half-past one o'clock, and a thousand dollars to raise; but there was no time for regrets. I started forth with a troubled heart, and not feeling very sanguine of success. Borrowing money is far from being pleasant employment, and is only endurable as a less evil than not meeting your obligations. For that day, I had thought my trials on this head over; but I erred. I had again to put on my armour of *brass* and go forth to meet coldness, rebuffs, and polite denials. Alas! I got no more; not a dollar rewarded my earnest efforts. Two o'clock found me utterly discouraged. Then, for the first time, it occurred to me to go to the holders of the note and frankly tell them that I could not lift it.

" But that will ruin your credit with them."

Yes, that was the rub; and then it was so mortifying a resource. After a short space of hurried reflection, I concluded that as I had twice as much credit in other quarters as it was prudent to use, I would ask a renewal of the note, which would be a great relief. It was better, certainly, than to suffer a protest. At the thought of a protest I shuddered, and started to see the parties to whom the note was due, feeling much as I suppose a culprit feels when about being arraigned for trial. It was twenty minutes past two when I called at their store.

"I am sorry," I said to one of the firm, whom I first met, speaking in a husky, agitated voice, "to inform you that I shall not be able to lift my note that falls due to-day."

His brows fell instantly.

"I had made every arrangement to meet it," I continued, "and was to receive the money at one o'clock to-day, but was unexpectedly disappointed. I have tried since to raise the amount, but find it too late in the day."

The man's brows fell still lower, while his eyes remained steadily fixed upon my face.

"I shall have to ask you to extend it for me."

"I don't think we can do that," he coldly replied.

"Will you consult your partners?" I said; "time presses."

The man bowed stiffly, his aspect about as pleasing as if I had robbed him, and turned away. I was standing near the door of the counting-room, inside of which were his two partners, with whom he had retired to confer.

"Jones can't pay his note," I heard him say, in tones most unpleasant to my ear.

"What!" was replied; "Jones?"

"Yes, Jones."

"What does he want?"

"A renewal."

"Nonsense! He can pay, if he finds he must."

"It is nearly half-past two," one of them remarked.

"No matter. It's of too much importance to him to keep his good name; he'll find somebody to help him. Threaten him with a protest; shake that over his head, and the money'll be raised."

With a Siberian aspect, the man returned to me.

"Can't do any thing for you," he said. "Sorry for it."

"My note must lie over, then," I replied

"It will be protested."

The very sound of the word went through me like an arrow. I felt the perspiration starting from every pore; but I was indignant at the same time, and answered, as firmly as I could speak—"Very well; let it be."

"As you like," he said, in the same cold tone, and with the same dark aspect, partly turning away as he spoke.

"But, my dear sir"—

"It is useless to waste words," he remarked, interrupting me. "You have our ultimatum."

As I left the store, I felt as if I had been guilty of some crime; I was ashamed to look even the clerks in the face. A feeble resolution to make an effort to save myself from the disgrace and disaster of a protest stirred in my mind; but it died away, and I returned to my store to await the dread result that must follow this failure to take up my paper. I looked at the slow-moving hand on the clock, and saw minute after minute go by with a stoicism that surprised even myself. At last the stroke of the hammer fell; the die was cast. I would be protested, that greatest of all evils dreaded by a man of business. As to going home to dinner, that was out of the question; I could not have eaten a mouthful to save me. All I had now to do was to wait for the visit of the notary, from which I shrank with a nervous dread. Everybody in the street would know him, I thought, and everybody would see him enter my store and comprehend his business.

Half-past three arrived, and yet I had not been bearded by the dread monster, at whose very name thousands have trembled and do still tremble. I sat awaiting him in stern silence. Four o'clock, and yet he had not come. Perhaps, it was suggested to me, the holders of the note had withdrawn it at the last moment. Cheering thought!

Just then I saw a lad enter the store and speak

to one of the clerks, who pointed back to where I sat. The boy was not over fourteen, and had, I noticed as he approached, a modest, rather shrinking look.

"Mr. Jones?" he said, when he had come near to me.

"Yes," I replied, indifferently, scarcely wondering what he wanted.

"Will you pay this note?" he said, opening a piece of paper that I had not observed in his hand, and presenting it to me.

My head was in a whirl for an instant, but was as quickly clear again.

"No, my lad," I replied, in a composed voice, "I shall not pay it."

"You will not pay it?" he repeated, as if he had not heard me distinctly.

"No," said I.

The lad bowed politely, slipped the dishonoured note into his pocket, and retired.

I drew a long breath, leaned back in my chair with a sense of relief, and murmured—"Not such a dreadful affair, after all. So, I am protested! The operation is over, and I hardly felt the pain. And now what next?"

As I said this, the man whose Siberian face had almost congealed me entered my store, and came hurriedly back to where I still remained sitting. His face was far less wintry. The fact was, I owed the firm fifteen thousand dollars, which was no joke; and they were nearly as much alarmed, when they found that my note was actually under protest, as I was before the fact.

"Is it possible, Mr. Jones," he said, his voice as husky and tremulous as mine was when I called upon him an hour or two before, "that you have suffered your note to lie over!"

"Did I not inform you that such would be the case?" I replied, with assumed sternness of voice and manner. The boot was on the other leg, and I was not slow in recognising the fact.

"But what do you intend to do, Mr. Jones? What is the state of your affairs?"

"At the proper time, I will inform you," I answered, coldly. "You have driven me into a protest, and you must stand the consequences."

"Are your affairs desperate, Mr. Jones?" The creditor became almost imploring in his manner.

"They will probably become so now. Does a man's note lie over without his affairs becoming desperate?"

"Perhaps"—

There was a pause. I looked unflinchingly into the man's face.

"If we extend this note, and keep the matter quiet, what then?"

"It won't do," I returned. "More than that will be required to save me."

My creditor looked frightened, while I maintained an aspect of as much indifference and resolution as I could assume.

"What will save you?" he asked.

I was thinking as rapidly as I could, in order to be prepared for striking while the iron was hot, and that to good purpose.

"I'll tell you," I replied.

"Well, what is it?" He looked eager and anxious.

"My fault has been one into which your house led me, that of buying too freely," said I; "of using my credit injudiciously. The consequence is, that I am cramped severely, and am neglecting my legitimate business in order to run about after money. I owe your house more than half of the aggregate of my whole liabilities. Give me the time I ask, in order to recover myself and curtail my business, and I can go through."

"What time do you ask?"

"I owe you fifteen thousand dollars."

"So much?"

"Yes; and the whole of it falls due within seven months. What I propose is, to pay you five per cent. on the amount of my present indebtedness every thirty days from this time until the whole is liquidated; you to hand me a thousand dollars to-morrow morning, to enable me to get my note out of bank, in order to save my credit."

The gentleman looked blank at the boldness of my proposition.

"Is that the best you can do?" he asked.

"The very best. You have driven me into a protest, and now, the bitterness of that dreaded ordeal being past, I prefer making an assignment and having my affairs settled up, to going on in the old way. I will not continue in business, unless I can conduct it easily and safely. I am sick of being on the rack; I would rather grub for a living."

I was eloquent in my tone and manner, for I felt what I said.

"It shall be as you wish," said my creditor. "You should not, you must not, make an assignment; every interest will suffer in that event. We will send you a check for a thousand dollars early to-morrow morning, and, as to what has occurred, keep our own counsel."

I bowed, and he bowed. I was conscious of having risen in his estimation. Get such a man in your power, and his respect for you increases fourfold.

My sleep was sound that night, for I was satisfied that the thousand dollars would come. And they did come.

After that, I was as easy as an old shoe. I was soon off the borrowing list; my business I contracted into a narrower and safer sphere, and really made more profit than before.

I have never stood in fear of notaries or protests since. Why should I? To me the notary proved a lamb rather than a lion, and my credit, instead of being ruined, was saved by a protest.

RETRENCHMENT;

OR, WHAT A MAN SAVED BY STOPPING HIS NEWSPAPER.

NOT many years ago, a farmer who lived a hundred or two miles from the seaboard, became impressed with the idea that unless he adopted a close-cutting system of retrenchment, he would certainly go to the wall. Wheat, during the preceding season, had been at a high price; but, unluckily for him, he had only a small portion of his land in wheat. Of corn and potatoes he had raised more than the usual quantity; but the price of corn was down, and potatoes were low. This year he had sown double the wheat he had ever sown before, and, instead of raising a thousand bushels of potatoes, as he had generally done, only planted about an acre in that vegetable, the product of which was about one hundred and fifty bushels.

Unluckily for Mr. Ashburn, his calculations did not turn out well. After his wheat was harvested, and his potatoes nearly ready to dig, the price of the former fell to ninety cents per bushel, and the price of the latter rose to one dollar. Everywhere, the wheat crop had been abundant, and almost everywhere the potato crop promised to be light.

Mr. Ashburn was sadly disappointed at this result.

"I shall be ruined," he said at home, and carried a long face while abroad. When his wife and daughters asked for money with which to get their fall and winter clothing, he grumbled sadly, gave them half what they wanted, and said they must retrench. A day or two afterwards, the collector of the "Post" came along and presented his bill.

Ashburn paid it in a slow, reluctant manner, and then said—

"I wish you to have the paper stopped, Mr. Collector."

"Oh, no, don't say that, Mr. Ashburn. You are one of our old subscribers, and we can't think of parting with you."

"Sorry to give up the paper. But must do it," returned the farmer.

"Isn't it as good as ever? You used to say you'd rather give up a dinner a week than the 'Post.'"

"Oh, yes, it's as good as ever, and sometimes I

think much better than it was. It's a great pleasure to read it. But I must retrench at every point, and then I don't see how I'm to get along. Wheat's down to ninety cents, and falling daily."

"But the paper is only two dollars a year, Mr. Ashburn."

"I know. But two dollars are two dollars. However, it's no use to talk, Mr. Collector; the 'Post' must be stopped. If I have better luck next year, I will subscribe for it again."

This left the collector nothing to urge, and he withdrew. In his next letter to the publishers, he ordered the paper to be discontinued, which was accordingly done.

Of this little act of retrenchment, Jane, Margaret, and Phœbe knew nothing at the time, and the far mer was rather loathe to tell them. When the fact did become known, as it must soon, he expected a buzzing in the hive, and the anticipation of this made him half repent of what he had done, and almost wish that the collector would forget to notify the office of his wish to have the paper stopped. But, the collector was a prompt man. On the second Saturday morning, Ashburn went to the post-office as usual. The postmaster handed him a letter saying, as he did so—

"I cant't find any paper for you, to-day. They have made a mistake in not mailing it this week."

"No," replied Ashburn. "I have stopped it."

"Indeed! The Post is an excellent paper. What other one do you intend to take?"

"I shall not take any newspaper this year," replied Ashburn.

"Not take a newspaper, Mr. Ashburn!" said the postmaster, with a look and in a tone of surprise.

"No. I must retrench. I must cut off all superfluous expenses. And I believe I can do without a newspaper as well as any thing else. It's a mere luxury; though a very pleasant one, I own, but still dispensable."

"Not a luxury, but a necessary, I say, and indispensable," returned the postmaster. "I don't know what I wouldn't rather do without than a newspaper. What in the world are Phœbe, and Jane, and Margaret going to do?"

"They will have to do without. There is no help for it."

"If they don't raise a storm about your ears that you will be glad to allay, even at the cost of half a dozen newspapers, I am mistaken," said the postmaster, laughing.

Ashburn replied, as he turned to walk away, that he thought he could face all storms of that kind without flinching.

"Give me the 'Post,' papa," said Margaret, running to the door to meet her father when she saw him coming.

"I haven't got it," replied Mr. Ashburn, feeling rather uncomfortable.

"Why? Hasn't it come?"

"No; is hasn't come."

Margaret looked very much disappointed.

"It has never missed before," she said, looking earnestly at her father.

No suspicion of the truth was in her mind; but, to the eyes of her father, her countenance was full of suspicion. Still, he had not the courage to confess what he had done.

"The 'Post' hasn't come!" he heard Margaret say to her sisters, a few minutes afterwards, and their expressions of disappointment fell rebukingly upon his ears.

It seemed to Mr. Ashburn that he heard of little else, while in the house, during the whole day, but the failure of the newspaper. When night came, even he, as he sat with nothing to do but think about the low price of wheat for an hour before bedtime, missed his old friend with the welcome face, that had so often amused, instructed, and interested him.

On Monday morning the girls were very urgent for their father to ride over to the post-office and see if the paper hadn't come; but, of course, the farmer was "too busy" for that. On Tuesday and Wednesday, the same excuse was made. On Thursday, Margaret asked a neighbour, who was going by the office, to call and get the newspaper for them. Towards evening, Mr. Markland, the neighbour, was seen riding down the road, and Margaret and Jane ran down eagerly to the gate for the newspaper.

"Did you get the paper for us?" asked Margaret, showing two smiling rows of milk-white teeth, while her eyes danced with anticipated pleasure.

Mr. Markland shook his head.

"Why?" asked both the girls at once.

"The postmaster says it has been stopped."

"Stopped!" How changed were their faces and tones of voice.

"Yes. He says your father directed it to be stopped."

"That must be a mistake," said Margaret. "He would have told us."

Mr. Markland rode on, and the girls ran back into the house.

"Father, the postmaster says you have stopped the newspaper!" exclaimed his daughters, breaking in upon Mr. Ashburn's no very pleasant reflections on the low price of wheat, and the difference in the return he would receive at ninety cents a bushel to what he would have realized at the last year's price of a dollar twenty-five.

"It's true," he replied, trenching himself behind a firm, decided manner.

"But why did you stop it, father?" inquired the girls.

"Because I can't afford to take it. It's as much

as I shall be able to do to get you enough to eat and wear this year."

Mr. Ashburn's manner was decided, and his voice had a repelling tone.

Margaret and Phœbe could say no more; but they did not leave their father's presence without giving his eyes the benefit of seeing a free gush of tears. It would be doing injustice to Mr. Ashburn's state of mind to say that he felt very comfortable, or had done so, since stopping the "Post," an act foı which he had sundry times more than half repented. But, as it had been done, he could not think of recalling it.

Very sober were the faces that surrounded the supper-table that evening; and but few words were spoken. Mr. Ashburn felt oppressed, and also fretted to think that his daughters should make both themselves and him unhappy about the trifle of a newspaper, when he had such serious troubles to bear.

On the next Saturday, as Mr. Ashburn was walking over his farm, he saw a man sitting on one of his fences, dressed in a jockey-cap, and wearing a short hunting-coat. He had a rifle over his shoulder, and carried a powder-flask, shot and bird bags. In fact, he was a fully equipped sportsman, a somewhat *rara avis* in those parts.

"What's this lazy fellow doing here?" said Ashburn, to himself. "I wonder where he comes from?"

"Good morning, neighbour," spoke out the stranger, in a familiar way, as soon as the farmer came within speaking distance. "Is there any good game about here? Any wild-turkeys, or pheasants?"

"There are plenty of squirrels," returned Ashburn, a little sarcastically, "and the woods are full of robbins."

"Squirrels make a first-rate pie. But I needn't tell you that, my friend. Every farmer knows the taste of squirrels," said the sportsman with great good-humour. "Still, I want to try my hand at a wild-turkey. I've come off here into the country to have a crack at game better worth the shooting than we get in the neighbourhood of P——."

"You're from P——, then?" said the farmer.

"Yes, I live in P——."

"When did you leave there?"

"Four or five weeks ago."

"Then you don't know what wheat is selling for now?"

"Wheat? No. I think it was ninety-five or a dollar, I don't remember which, when I left."

"Ninety is all it is selling for here."

"Ninety! I should like to buy some at that."

"I have no doubt you can be accommodated," replied the farmer.

"That is exceedingly low for wheat. If it wasn't for having a week's sport among your wild-turkeys and the hope of being able to kill a deer, I'd stop and buy up a lot of wheat on speculation."

"I'll sell you five hundred bushels at ninety-two," said the farmer, half-hoping that this green customer might be tempted to buy at this advance upon the regular rate.

"Will you?" interrogated the stranger.

"Yes."

"I'm half-tempted to take you up. I really believe I—no!—I must knock over some wild-turkeys first. It won't do to come this far without bagging rarer game than wheat. I believe I must decline, friend."

"What would you say to ninety-one?" The farmer had heard a rumour, a day or two before, of a fall of two or three cents in wheat, and if he could get off five hundred bushels upon this sportsman, who had let the breast of his coat fly open far enough to give a glimpse of a large, thick pocket-book, at ninety-one, it would be quite a desirable operation.

"Ninety-one—ninety-one," said the stranger, to himself. "That is a temptation! I can turn a penny on that. But the wild-turkeys; I must have a crack at a wild-turkey or a deer. I think, friend," he added, speaking louder, "that I will have some sport in these parts for a few days first. Then, maybe, I'll buy up a few thousand bushels of wheat, if the prices haven't gone up."

"I shouldn't wonder if prices advanced a little," said the farmer.

"Wouldn't you?" And the stranger looked into the farmer's face with a very innocent expression.

"It can't go much lower; if there should be any change, it will doubtless be an improvement."

"How much wheat have you?" asked the sportsman.

"I've about a thousand bushels left."

"A thousand bushels. Ninety cents; nine hundred dollars;—I'll tell you what, friend, since talking to you has put me into the notion of trying my hand at a speculation on wheat, I'll just make you an offer, which you may accept or not, just as you please. I'll give you ninety cents cash for all you've got, one half payable now, and the other half on delivery of the wheat at the canal, provided you get extra force and deliver it immediately."

Ashburn stood thoughtful for a moment or two and then replied—

"Very well, sir, it's a bargain."

"Which, to save time, we will close immediately I will go with you to your house, and pay you five hundred dollars on the whole bill for a thousand bushels."

The farmer had no objection to this, of course, and invited the stranger to go to his house with him, where the five hundred dollars were soon counted out. For this amount of money he wrote a receipt and handed it to the stranger, who, after reading it, said—

"I would prefer your making out a bill for a thousand bushels, and writing on it, 'Received on account, five hundred dollars.'"

"It may overrun that quantity," said Ashburn.

"No matter, a new bill can be made out for that. I'll take all you have."

The farmer saw no objection to the form proposed by the stranger, and therefore tore up the receipt he had written, and made a bill out in the form desired.

"Will you commence delivering to-day?" inquired the sportsman, who all at once began to manifest a marked degree of interest in the business.

"Yes," replied the farmer.

"How many wagons have you?"

"Two."

"As it is down hill all the way to the canal, they can easily take a hundred bushels each."

"Oh, yes."

"Very well. They can make two loads apiece to-day, and, by starting early, three loads apiece on Monday, which will transfer the whole thousand bushels to the canal. I will go down immediately and see that a boat is ready to commence loading. You can go to work at once."

By extra effort, the wheat was all delivered by Monday afternoon, and the balance of the purchase money paid. As Mr. Ashburn was riding home, a neighbour who had noticed his wagons going past his house with wheat for the two days, overtook him.

"So I see, friend Ashburn, that, like me, you are content to take the first advance of the market, instead of running the risk of a decline for a further rise in prices. What did you get for your wheat?"

"I sold for ninety cents."

"Ninety cents!" exclaimed the neighbour. "Surely you didn't sell for that?"

"I certainly did. I tried to get ninety-two, but ninety was the highest offer I could obtain."

"Ninety cents! Why, what has come over you, Ashburn. Wheat is selling for a dollar and twenty cents. I've just sold five hundred bushels for that."

"Impossible!" ejaculated the farmer.

"Not at all impossible. Don't you know that by the last arrival from England have come accounts of a bad harvest, and that wheat has taken a sudden rise?"

"No, I don't know any such a thing," returned the astonished Ashburn.

"Well, it's so. Where is your newspaper?—Haven't you read it? I got mine on Friday evening, and saw the news. Early on Saturday morning I found two or three speculators ready to buy up all the wheat they could get at old prices; but they didn't make many operations. One fellow, who pretended to be a fancy sportsman, thrust himself into my way, but, even if I had not known of a rise in the price of wheat, I should have suspected it as soon as I saw him, for I read, last week, of just such a looking chap as him having got ahead of some ignorant country farmers by buying up their produce, on a sudden rise of the market, at a price much below its real value."

"Good day!" said Ashburn, suddenly applying his whip to the flank of his horse; and away he dashed homeward at a full gallop.

The farmer never sat down to make a regular calculation of what he had lost by stopping his newspaper; but it required no formality of pencil and paper to arrive at this. A difference of thirty cents on each bushel, made, for a thousand bushels, the important sum of three hundred dollars, and this fact his mind instantly saw.

By the next mail, he enclosed two dollars to the publishers of the "Post," and re-ordered the paper. He will, doubtless, think a good while, and retrench at a good many points, before he orders another discontinuance.

HUNTING UP A TESTIMONIAL.

"DOCTOR," said a man with a thin, sallow countenance, pale lips, and leaden eyes, coming up to the counter of a drug-store in Baltimore, some ten years ago—"Doctor, I've been reading your advertisement about the 'UNIVERSAL RESTORER, AND BALSAM OF LIFE,' and if that Mr. John Johnson's testimony is to be relied on, it ought to suit my case, for, in describing his own sufferings, he has exactly described mine. But I've spent so much money in medicine, to no purpose, that I am tired of being humbugged: so, if you'll just tell me where I can find this Mr. Johnson, I'll give him a call. I'd like to know if he's a real flesh-and-blood man."

"You don't mean to insinuate that I'd forge a testimonial?" replied the man of medicine, with some slight show of indignation.

"Oh, no. I don't insinuate any thing at all, doctor," answered the pale-looking man. "But I'd like to see this Mr. John Johnson, and have a little talk with him."

"You can do that, if you'll take the trouble to call on him," said the doctor, in an off-hand way.

"Where can I find him?" asked the man.

"He lives a little way out of town; about three miles on the Fredrick turnpike."

"Ah, so far?"

"Yes. Go out until you come to the three-mile stone; then keep on to the first road, turning off to the right, along which you will go about a quarter of a mile, when you will see a brick house. Mr. Johnson lives there."

The thin, sallow-faced man bowed and retired. As he left the store, the doctor gave a low chuckle, and then said, half aloud—"I guess he won't try to find this Mr. John Johnson."

But he was mistaken. Three hours afterwards, the sick man entered the shop, and, sinking upon a chair with an expression of weariness, said, in a fretful tone—

"Well, doctor, I've been out where you said, but no Mr. John Johnson lives there."

"Mr. Johnson lives at the place to which I directed you," said the doctor, positively.

But the man shook his head.

"You went out the Fredrick road to the three-mile stone?"

"Yes."

"And turned off at the first road on the left-hand side?"

"You told me the *right* hand side!" said the man.

"Oh, there's the mistake," replied the doctor, with the air of a man who had discovered a very material error, by which an important result was affected; "I told you to turn off to the *left*."

"I'm sure you said the right," persisted the man.

"Impossible!" returned the doctor, in a most confident tone of voice. "How could I have said the right-hand side when I knew it was the left? I know Mr. Johnson as well as I know my own brother, and have been at his house hundreds of times."

"I am almost sure you said the right!" persisted the man.

"Oh, no! You misunderstood me," most positively answered the doctor.

"Well, I must only try it again," said the man, languidly; "but shall have to defer the walk until to-morrow, for I'm completely worn down."

"You'd better try a bottle of the RESTORER," said the doctor with a benevolent smile. "I know it will just suit your case. Mr. Johnson looked worse than you do, when he commenced taking it, and three bottles made a well man of him."

And the doctor held up a bottle of the Restorer, with its handsome label, temptingly, before the eyes of the sick man, adding, as he did so—

"It is only fifty cents."

"I've been humbugged too often!" replied the suspicious patron of patent-medicine venders. "No; I'll see Mr. Johnson first."

"Well, did you see Mr. Johnson?" asked the doctor with a pleasant smile and confident air, as the testimonial-hunter entered his shop on the next day, about noon.

"No, I did not," was replied, a little impatiently.

"Ah? How comes that? Did you follow the directions I gave?"

"Yes, to the very letter."

"Then you must have found Mr. Johnson."

"But I tell you, I didn't."

"It's very strange! I can't understand it. You turned off at the first road to the left, after passing the third milestone?"

"I did."

"Two tall poplars stood at the gate which opened from the turnpike?"

"What gate?

"The gate opening into the lane leading to Mr. Johnson's house."

"I didn't turn of at any gate," said the man. "I kept on, as you directed, to the first road that led off from the turnpike. You didn't mention any thing about a gate."

"I didn't suppose it necessary," replied the doctor, with a show of impatience. "A road is a road, whether you enter it by a gate or in any other manner. Roads leading to gentlemen's country-seats are not usually left open for every sort of ingress and egress. I don't wonder that you were unable to find Mr. Johnson."

"I wish you'd give me a more particular direction," said the invalid. "I'm nearly dead now with fatigue; I'll try once more to find this man, and if I don't turn him up, I'll let the matter drop. I don't believe your medicine will do me much good, anyhow."

"I'm sure it will help you," replied the doctor. "I can tell from your very countenance that it is what you want. Hundreds affected as you are have been restored to health. Better take a bottle."

"I want to see this Mr. Johnson first," persisted the sick man.

"Get a carriage, then. This walking in the hot sun is too much for you."

"Can't afford to ride in carriages. Have spent all my money in doctor-stuffs. Oh, dear! Well! You say this man lives just beyond the three-mile stone, at the first road leading off to the left?"

"Yes."

"Two poplars stand at the gate?"

"Yes."

"I ought to find that," said the man.

"You can find it, if you try," returned the doctor.

The man started off again.

"Plague on the persevering fellow!" muttered the man of drugs, as soon as the invalid retired. "I wish I'd sent him six miles, instead of three."

The day wore on, but the testimonial-hunter did not reappear. Early on the next morning, however, his pale, thin face and emaciated brows were visible in the shop of the quack-doctor.

"Ah! good morning! good morning!" cried the latter, with one of the most assured smiles in the world. "You found Mr. Johnson, and pleasant of course?"

"Confound you, and Mr. Johnson, too! No!" replied the invalid impatiently.

The doctor was a man of great self-control, and, of course, did not in the least become offended.

"Strange!" said he, seriously. "You surely didn't follow my directions."

"I surely did. The first gate on the left-hand side. But your two tall poplars was one tall elm."

"There it is again!" and the doctor, in the fulness of his surprise, actually let a small package, that he held in his hand, fall upon the counter. "I told you poplars, distinctly. The elm-tree gate is at least a quarter of a mile this side. But, to settle the matter at once," and the doctor, speaking like a man who was about doing a desperate thing, turned to his shelves and took therefrom a bottle of the Universal Restorer—" here's the medicine. I know it will cure you. Take a bottle. It shall cost you nothing."

The sick man, tempted strongly by the hope of a cure, hesitated for a short time, and then said—

"I don't want your stuff for nothing. But half a dollar won't kill me."

So he drew a coin from his pocket, laid it upon the counter, and, taking the medicine, went slowly away.

"Rather a hard customer that," said the doctor to himself, with a chuckle, as he slipped the money in his drawer. "But I'll take good care to send the next one like him a little farther on his fool's errand. He'd much better have taken my word for it in the beginning."

The sick man never came back for a second bottle of the "Restorer." Whether the first bottle killed or cured him is, to the chronicler, unknown

TRYING TO BE A GENTLEMAN.

THE efforts which certain young men make, on entering the world, to become gentlemen, is not a little amusing to sober, thoughtful lookers on. To "become" is not, perhaps, what is aimed at, so much as to make people believe that they are gentlemen; for if you should happen to insinuate any thing to the contrary, no matter how wide from the mark they go, you may expect to receive summary punishment for your insolence.

One of these characters made himself quite conspicuous, in Baltimore, a few years ago. His name was L——, and he hailed from Richmond, we believe, and built some consequence upon the fact that he was a son of the Old Dominion. He dressed in the extreme of fashion; spent a good deal of time strutting up and down Market street, switching his rattan; boarded at one of the hotels; drank wines freely, and pretended to be quite a judge of their quality; swore round oaths occasionally, and talked of his honour as a gentleman.

His knowledge of etiquette he obtained from

books, and was often quite as literal in his observance of prescribing modes and forms, as was the Frenchman in showing off his skill in our idioms, when he informed a company of ladies, as an excuse for leaving them, that he had "some fish to fry." That he was no gentleman, internally or externally, was plain to every one; yet he verily believed himself to be one of the first water, and it was a matter of constant care to preserve the reputation.

Among those who were thrown into the society of this L——, was a young man, named Briarly, who had rather more basis to his character, and who, although he dressed well, and moved in good society, by no means founded thereon his claim to be called a gentleman. He never liked L——, because he saw that he had no principle whatever; that all about him was mere sham. The consequence was that he was hardly civil to him, a circumstance which L—— was slow either to notice or resent.

It happened, one day, that the tailor of Briarly asked him if he knew any thing about L——.

"Not much," replied Briarly. "Why do you ask?"

"Do you think him a gentleman?"

"How do you estimate a gentleman?" asked the young man.

"A gentleman is a man of honour," returned the tailor.

"Very well; then L—— must be a gentleman, for he has a great deal to say about his honour."

"I know he has; but I find that those who talk much of their honour, don't, as a general thing, possess much to brag of."

"Then, he talks to you of his honour?"

"Oh, yes; and gives me his word as a gentleman."

"Does he always keep his word as a gentleman?"

The tailor shrugged his shoulders.

"Not always," he replied.

"Then I should say that the word of a gentleman isn't worth much," smilingly remarked Briarly.

"Not the word of such broadcloth and buckram gentlemen as he is."

"Take care what you say, or you may find yourself called to account for using improper language about this gentleman. We may have a duel on the carpet."

"It would degrade him to fight with a tailor," replied the man of shears. "So I may speak my mind with impunity. But if he should challenge me, I will refuse to fight him, on the ground that he is no gentleman."

"Indeed! How will you prove that?"

"Every man must be permitted to have his own standard of gentility."

"Certainly."

"I have mine."

"Ah! Well, how do you measure gentility?"

"By my ledger. A man who doesn't pay his tailor's bill, I consider no gentleman. If L—— sends me a challenge, I will refuse to fight him on that ground."

"Good!" said Briarly, laughing. "I'm afraid, if your standard were adopted, that a great many, who now pass themselves off for gentlemen, would be held in little estimation."

"It is the true standard, nevertheless," replied Shears. "A man may try to be a gentleman as much as he pleases, but if he don't try to pay his tailor's bill at the same time, he tries in vain."

"You may be right enough," remarked Briarly, a good deal amused at the tailor's mode of estimating a gentleman, and possessed of a new fact in regard to L——'s claim to the honourable distinction of which he so often boasted.

Shortly after this, it happened that L—— made Briarly angry about something, when the latter very unceremoniously took hold of the handle on the young man's face, and moved his head around. Fortunately, the body moved with the head, or the consequences might have been serious. There were plenty to assure L—— that for this insult he must, if he wished to be considered a gentleman, challenge Briarly, and shoot him—if he could. Several days elapsed before L——'s courage rose high enough to enable him to send the deadly missive by the hand of a friend.

Meantime, a wag of a fellow, an intimate friend of Briarly's, appeared in Market street in an old rusty coat, worn hat, and well-mended but clean and whole trowsers and vest. Friend after friend stopped him, and, in astonishment, inquired the cause of this change. He had but one answer, in substance. But we will give his own account of the matter, as related to three or four young bucks in an oyster-house, where they happened to meet him. L—— was of the number.

"A patch on your elbow, Tom, as I live!" said one; "and here's another on your vest. Why, old fellow, this is premeditated poverty."

"Better wear patched garments than owe for new ones," replied Tom, with great sobriety.

"Bless us! when did you turn economist?"

"Ever since I tried to be a gentleman."

"What?"

"Ever since I tried to be a gentleman. I may strut up and down Market street in fine clothes, switch my rattan about, talk nonsense to silly ladies, swear, and drink wine; but if I don't pay my tailor, I'm no gentleman."

"Nonsense," was replied. There was a general laugh, but few of Tom's auditors felt very much flattered by his words.

"No nonsense at all," he said. "We may put

on airs of gentility, boast of independence and spirit, and all that; but it's a mean kind of gentility that will let a man flourish about in a fine coat for which he owes his tailor. Wyville has a large bill against me for clothes, Grafton another for boots, and Cox another for hats. I am trying to pay these off—trying to become a gentleman."

"Then you don't consider yourself a gentleman now?" said one.

"Oh, no. I'm only trying to become a gentleman," meekly replied Tom, though a close observer could see a slight twitching in the corner of his mouth, and a slight twinkle in the corner of his eye. "My honour is in pawn, and will remain so until I pay these bills. Then I shall feel like holding up my head again, and looking gentlemen in the face."

The oddness of this conceit, and the boldness with which it was carried out, attracted attention, and made a good deal of talk at the time. A great many tailors' bills were paid instanter that would not have been paid for months, perhaps not at all.

In a few days, however, Tom appeared abroad again, quite as handsomely dressed as before, alleging that his uncle had taken compassion on him, and, out of admiration for his honest principles, paid off his bills and made a gentleman of him once more.

No one, of course, believed Tom to be sincere in

all this. It was looked upon as one of his waggish tricks, intended to hit off some one, or perhaps the whole class of fine tailor-made gentlemen who forget their benefactors.

While Tom was metamorphosed as stated, Briarly was waited upon one day, by a young man, who presented him with a challenge to mortal combat from the insulted L——, and desired him to name his friend.

"I cannot accept the challenge," said Briarly, promptly.

"Why not?" asked the second of L——, in sur prise.

"Because your principal is no gentleman."

"What!"

"Is no gentleman," coolly returned Briarly.

"Explain yourself, sir, if you please."

"He doesn't pay his tailor, he doesn't pay his boot-maker, he doesn't pay his hatter—he is, therefore, no gentleman, and I cannot fight him."

"You will be posted as a coward," said the second, fiercely.

"In return for which I will post him as no gentleman, and give the evidence," replied Briarly.

"I will take his place. You will hear from me shortly," said the second, turning away.

"Be sure you don't owe your tailor any thing, for if you do, I will not stoop to accept your challenge," returned Briarly. "I will consider it *pri-*

mâ facie evidence that you are no gentleman. I know Patterson very well, and will, in the mean time, inform myself on the subject."

All this was said with the utmost gravity, and with a decision of tone and manner that left no doubt of the intention.

The second withdrew. An hour elapsed, but no new challenge came. Days went by, but no "posters" drew crowds at the corners. Gradually, the matter got wind, to the infinite amusement of such as happened to know L——, who was fairly driven from a city where it was no use trying to be a gentleman without paying his tailor's bill.

TAKING A PRESCRIPTION.

SUMMER before last, the time when cholera had poisoned the air, a gentleman of wealth, standing and intelligence, from one of the Southern or Middle States, while temporarily sojourning in Boston, felt certain "premonitory symptoms," that were rather alarming, all things considered. So he inquired of the hotel-keeper where he could find a good physician.

"One of your best," said he, with an emphasis in his tones that showed how important was the matter in his eyes.

"Doctor —— stands at the head of his profession in our city," returned the hotel-keeper. "You may safely trust yourself in his hands."

"Thank you. I will call upon him immediately," said the gentleman, and away he went.

The doctor, fortunately, as the gentleman mentally acknowledged, was in his office. The latter, after introducing himself, stated his case with some concern of manner; when the doctor felt his pulse, looked at his tongue, and made sundry professional inquiries.

"Your system is slightly disturbed," remarked the doctor, after fully ascertaining the condition of his patient, "but I'll give you a prescription that will bring all right again in less than twenty-four hours."

And so he took out his pencil and wrote a brief prescription.

"How much am I indebted, doctor?" inquired the gentleman, as he slipped the little piece of paper into his vest pocket.

"Five dollars for the consultation and prescription," replied the doctor, bowing.

"Cheap enough, if I am saved from an attack of cholera," said the patient as he drew forth his pocket-book and abstracted from its folds the required fee. He then returned to the hotel, and, going to one of the clerks, or bar-keeper, in the office, said to him—

"I wish you would send out and get me this prescription."

"Prescription! Why, Mr. ——, are you sick?" returned the bar-keeper.

"I'm not very well," was answered.

"What's the matter?"

"Symptoms of the prevailing epidemic."

"Oh! Ah! And you've been to see a doctor?"

"Yes."

"Who?"

"Doctor ——."

The bar-keeper shrugged his shoulders, as he replied—

"Good physician. None better. That all acknowledge. But, if you'll let me prescribe for you, I'll put you all straight in double-quick time."

"Well, what will you prescribe, Andy?" said the gentleman.

"I'll prescribe this." And, as he spoke, he drew from under the counter a bottle labelled—"Mrs. ——'s Cordial." "Take a glass of that, and you can throw your doctor's prescription into the fire."

"You speak confidently, Andy?"

"I do, for I know its virtue."

The gentleman, who had in his hand a prescription for which he had paid five dollars to one of the most skilful and judicious physicians in New England, strange as it may seem, listened to this bar-keeper, and in the end actually destroyed the prescription, and poured down his throat a glass of "Mrs. ——'s Cordial."

It is no matter of surprise that, ere ten o'clock in the evening, the gentleman's premonitory symptoms, which had experienced a temporary abatement, assumed a more alarming character. And now, instead of going to, he was obliged to send for, a physician. Doctor ——, whom he had consulted, was called in, and immediately recognised his patient of the morning.

"I'm sorry to find you worse," said he. "I did not in the least doubt the efficacy of the remedy I gave you. But, have you taken the prescription."

"Wh— wh— why no, doctor," stammered the half-ashamed patient. "I confess that I did not. I took something else."

"Something else! What was it?"

"I thought a glass of Mrs. ——'s cordial would answer just as well."

"You did! and, pray, who prescribed this for you?" said the doctor, moving his chair instinctively from his patient and speaking in a rather excited tone of voice.

"No one prescribed it. I took it on the recommendation of the bar-keeper down-stairs, who said that he knew it would cure me."

"And you had my prescription in your pocket at the same time! The prescription of a regular physician, of twenty-five years' practice, set aside for a quack nostrum, recommended by a bar-keeper! A fine compliment to common sense and the profession, truly! My friend, if I must speak out plainly, you deserve to die—and I shouldn't much wonder if you got your deserts! Good evening!"

Saying this, the doctor arose, and was moving towards the door, when the frightened patient called to him in such appealing tones, that he was constrained to pause. A humble confession of error, and repeated apologies, softened the physician's suddenly awakened anger, and he came back and resumed his seat.

"My friend," said he, on recovering his self-possession, which had been considerably disturbed, "Do you know the composition of Mrs. ——'s cordial, which you took with so much confidence?"

"I do not!" replied the gentleman.

"Humph! Well, I can tell you. About nine-tenths of it is cheap brandy, or New-England rum, which completely destroys or neutralizes the salutary medicaments that form the tithe thereof. I don't wonder that this stuff has aggravated all your symptoms. I would, if in your state of health, about as leave take poison"

"Pray, don't talk to me in that way, doctor," said the patient, imploringly. "I am sick, and what you say can only have the effect to make me worse. I am already sufficiently punished for my folly. Prescribe for me once more, and be assured that I will not again play the fool."

Doctor ——'s professional indignation had pretty well burned itself out by this time; so he took up the case again, and once more gave a prescription.

In a couple of days, the gentleman was quite well again; but that Mrs. ——'s cordial cost him twenty dollars.

He is now a little wiser than he was before; and is very careful as to whose prescriptions he takes. It would be better for the health of the entire community if every individual would be as careful in the same matter as he is now. Those who are sick should, ere taking medicine, consult a physician of experience and skill; but, above all things, they should shun advertised nostrums, in the sale of which the manufacturers and vendors are interested. Often testimonials as to their efficacy are mere forgeries. Health is too vital a thing to be risked in this way.

THE YANKEE AND THE DUTCHMAN; OR, I'LL GIVE OR TAKE.

A SHREWD Yankee, with about five hundred dollars in his pocket, came along down South, a few years ago, seeking for some better investment of his money than offered in the land of steady habits, where he found people, as a general thing, quite as wide awake as himself.

In Philadelphia, our adventurer did not stay long; but something in the air of Baltimore pleased him, and he lingered about there for several weeks, prying into every thing and getting acquainted with everybody that was accessible. Among others for whom the Yankee seemed to take a liking, was a Dutchman, who was engaged in manufacturing an article for which there was a very good demand, and on which there was a tempting profit. He used to drop in almost every day and have a talk with the Dutchman, who seemed like a good, easy kind of a man, and just the game for the Yankee, if he should think it worth the candle.

"Why don't you enlarge your business?" asked

Jonathan, one day. "You can sell five times what you make."

"I knows dat," returned the Dutchman, "but I wants de monish. Wait a while, den I enlarsh."

"Then you are laying by something?"

"Leetle mite."

In two or three days, Jonathan came round again. He had thought the matter all over, and was prepared to invest his five hundred dollars in the Dutchman's business, provided the latter had no objections.

"It's a pity to creep along in the way you are going," he said, "when so much money might be made in your business by the investment of more capital. Can't you borrow a few hundred dollars?"

"Me borrow? Oh, no; nobody lend me few hunnard dollar. I go on, save up; bimeby I enlarsh."

"But somebody else, with plenty of money, might go into the business and fill the market; then it would be no use to enlarge."

"Sorry, but can't help it. No monish, no enlarsh."

"I've got five hundred dollars."

The phlegmatic Dutchman brightened up

"Fife hunnard dollar?"

"Yes."

"Much monish. Do great business on fife hunnard dollar."

"That you could."

"You lend me de monish?" asked the Dutchman.

Jonathan shook his head.

"Can't do that. I'm going into business myself."

"Ah! what business?"

"Don't know yet; haven't decided. Into your business, maybe."

"My business!" The Dutchman looked surprised.

"Yes; it appears to me like a very good business. Don't you think I could start very fair on five hundred dollars?"

The Dutchman hesitated to answer that question; he didn't want to say yes, and he was conscious that the Yankee knew too much of his affairs to believe him if he said no. He, therefore, merely shrugged his shoulders, looked stupid, and remained silent.

"You don't know of a large room that I could get anywhere, do you?"

The Dutchman shook his head, and gave a decided negative.

Jonathan said no more on that occasion. Two days afterwards, he dropped in again.

"Have you fount a room yet?" asked the Dutchman.

"I've seen two or three," replied Jonathan "One of them will suit me, I guess. But I'll tell you what I've been thinking about since I saw you. If I open another establishment, the business will be divided. Now, it has struck me, that, perhaps, it might be better, all round, for me to put my five hundred dollars into your business as a partner, and push the whole thing with might and main. How does it strike you?"

"Vell, I can't say shust now; I'll dink of him. You put in fife hunnard dollar, you say?"

"Yes; five hundred down, in hard cash—every dollar in gold."

"Fife hunnard. Let us see." And the Dutchman raised his chin and dropped his eyes, and stood for some minutes in a deep study.

"Fife hunnard," he repeated several times.

"Come to-morrow," he at length said. "Den I tell you."

"Very well. I'll drop in to-morrow," replied the Yankee. "I'm not very anxious about it, you see; but, as the thing occurred to me, I thought I would mention it. Five hundred dollars will make a great difference in your business."

On the next day, Jonathan appeared, looking quite indifferent about the matter. The Dutchman had turned over the proposition, and dreamed about it, both sleeping and waking. His final decision was to take in the Yankee as a partner.

Now, a cool, thoughtful Dutchman, and a quick-witted Yankee, are not a very bad match for each other, provided the former sees reason to have his wits about him, which was the case in the present instance. The Dutchman meant all fair; he had no thought of taking any advantage: but he had suspicion enough of Jonathan to put him on his guard, and look to see that no high-handed game was played off upon him.

"You put in fife hunnard dollar?" he said, when the Yankee appeared.

"Yes."

"Hard cash?"

"Yes, in gold."

"Gold!"

"All in half-eagles like these." And he drew a handful of gold coins from his pocket.

"Very well; I dake you. You put in fife hunnard dollar, I put in all I got here; den we joint owner."

"Equal partners?"

"Yes."

"That is, I own half and you half."

"Yes."

"And we divide, equally, the profits?"

"Yes."

"Very well; that'll do, I guess. We'll have writings drawn to this effect—articles of co-partnership, you know."

"Oh, yes."

This settled, nothing remained but to have the articles drawn, the money paid in, and the agreement signed and witnessed; all of which was done in the course of a few weeks. Then Jonathan went into the business, and infused some Yankee spirit into every part of it; he made things move ahead fast. In less than a year, the business was much more than doubled, and the profits in proportion; but Jonathan was not satisfied with his half of these—he wanted the whole; and, hedge-hog-like, he did all he could, by merely bristling up, to make things unpleasant for his partner. But the Dutchman was by no means thin-skinned; the sharp spikes of the Yankee's character annoyed him but little. As for himself, he felt very well satisfied with his share of the profits, and willing to go on as they were going.

At the end of the second year, when the establishment had grown into quite an important and profitable concern, the Yankee had a visit from an Eastern friend, a man of some capital.

"That's a stupid-looking fellow, that partner of yours," said this person.

"And he is as stupid as a mule. I have to carry him on my back, and the business, too."

"Why don't you get rid of him?"

"I've been wanting to do so for some time, but haven't seen my way clear yet."

"Does your partnership expire at any time, by limitation?"

"No. It can only be dissolved by mutual consent."

"Won't he sell out his interest?"

"I don't know; but I've always intended to make him an offer to give or take, as soon as I could see my way clear to do it."

"Don't you see your way clear now?"

"No. When such an offer is made, it must be of a sum that it is impossible for him to raise; otherwise, he might agree to give the amount proposed, and I don't want that. I wish to stick to the business, for it's going to be a fortune. At present, I am not able to raise what I think should be offered."

"How much is that?"

"About three thousand dollars. I only put in five hundred, two years ago. You can see how the business has increased. The half is worth five thousand in reality, and I would give, rather than take that sum."

"You think your partner can't raise three thousand dollars?"

"Oh, no; he's got no friends, and he hasn't three hundred out of the business."

"How long would you want the sum mentioned?"

"A year or eighteen months."

"I reckon I can supply it," said the friend. "It's a pity for you to be tied to this old Dutchman, when you can conduct the business just as well yourself."

"A great deal better; he is only in my way."

"Very well. You make him the offer to give or take three thousand dollars, and I will supply the money. But you ought, by all means, to add a stipulation, that whoever goes out shall sign a written agreement not to go into the same business for at least ten years to come. If you don't do this, he can take his three thousand dollars and start another establishment upon as large a scale as the one you have, and seriously affect your operations."

"Such a stipulation must be signed, of course," remarked Jonathan. "I've always had that in my mind; let me once get this business into my hands, and I'll make it pay better than it ever has yet. Before ten years roll over my head, if I a'n't worth forty or fifty thousand dollars, then I don't know any thing."

"You think it will pay like that?"

"Yes, I know it. I haven't put out half my strength yet, for I didn't want to let this Dutchman see what could be made of the business. He'll catch at three thousand dollars like a trout at a fly; it's more money than he ever saw in his life."

On the next day, Jonathan told his partner that he wanted to have some talk with him; so they retired into their little private office, to be alone.

"Vat you want?" said the Dutchman, when they were by themselves; for he saw that his partner had something on his mind of graver import than usual.

"I'm tired of a co-partnership business," said the Yankee, coming straight to the main point.

"Vell?" And the Dutchman looked at him without betraying the least surprise.

"Either of us could conduct this business as well as both together."

"Vell?"

"Now, I propose to buy you out or sell you my interest, as you please."

"Vell?"

"What will you give me for my half of the business, and let me go at something else?"

The Dutchman shook his head.

"At a word, then, to make the matter as simple as possible, and as fair as possible, I'll tell you what I'll give or take."

"Vell?"

"Of course, it would not be fair for the one who goes out to commence the same business. I would not do it. There should be a written agreement to this effect."

"Yes. Vell, vat vill you give or dake?"

"I'll give or take three thousand dollars; I don't care which."

"Dree dousand dollar! You give dat?"

"Yes."

"Or take dat?"

"Either."

"You pay down de monish?"

"Cash down."

"Humph! Dree dousand dollar! Me tink about him."

"How long do you want to think?"

"Undil de mornin."

"Very well; we'll settle the matter to-morrow morning."

In the morning, Jonathan's friend came with three thousand dollars, in order to pay the Dutchman right down, and have the whole business concluded while the matter was warm.

Meantime, the Dutchman, who was not quite so friendless nor so stupid as the Yankee supposed, turned the matter over in his mind very coolly. He understood Jonathan's drift as clearly as he understood it himself, and was fully as well satisfied as he was in regard to the future value of the business which he had founded. Two of their largest customers were Germans, and to them he went and made a full statement of his position, and gave them evidence that entirely satisfied them as to the business. Without hesitation, they agreed to advance him the money he wanted, and to enable him to strike while the iron was hot, checked him out the money on the next morning. One of them accompanied him to his manufactory, to be a witness in the transaction.

Jonathan and his friend were first on the spot.

In about ten minutes, the Dutchman and his friend arrived.

"Well, have you made up your mind yet?" asked the Yankee.

"De one who goes out ish not to begin de same business?"

"No, certainly not; it wouldn't be fair."

"No, I 'spose not."

"Suppose we draw up a paper, and sign it to that effect, before we go any farther."

"Vell."

The paper was drawn, signed, and witnessed by the friends of both parties.

"You are prepared to give or take?" said Jonathan, with some eagerness in his manner.

"Yes."

"Well, which will you do?"

"I vill give," coolly replied the Dutchman.

"Give!" echoed the Yankee, taken entirely by surprise at so unexpected a reply. "Give! You mean, take."

"I no means dake, I means give. Here ish de monish;" and he drew forth a large roll of bank-bills. "You say give or dake—I say give."

With the best face it was possible to put upon the matter, Jonathan, who could not back out, took the three thousand dollars, and, for that sum, signed away, on the spot, all right, title, and claim to benefit in the business, from that day henceforth and for ever.

With his three thousand dollars in his pocket, the Yankee started off farther South, vowing that, if he lived to be as old as Methuselah, he'd never have any thing to do with a Dutchman again.

A TIPSY PARSON.

In a village not a hundred miles from Philadelphia, resided the Rev. Mr. Manlius, who had the pastoral charge of a very respectable congregation, and was highly esteemed by them; but there was one thing in which he did not give general satisfaction, and in consequence of which many excellent members of his church felt seriously scandalized. He would neither join a temperance society, nor omit his glass of wine when he felt inclined to take it. It is only fair to say, however, that such spirituous indulgences were not of frequent occurrence It was more the principle of the thing, as he said, that he stood upon, than any thing else, that prevented his signing a temperance pledge.

Sundry were the attacks, both open and secret,

to which the Reverend Mr. Manlius was subjected, and many were the discussions into which he was drawn by the advocates of total abstinence. His mode of argument was very summary.

"I would no more sign a pledge not to drink brandy than I would sign a pledge not to steal," was the position he took. "I wish to be free to choose good or evil, and to act right because it is wrong to do otherwise. I do not find fault with others for signing a pledge, nor for abstaining from wine. If they think it right, it is right for them. But as for myself, I would cut off my right hand before I would bind myself by mere external restraint. My bonds are internal principles. I am temperate because intemperance is sin. For men who have abused their freedom, and so far lost all rational control over themselves that they cannot resist the insane spirit of intemperance, the pledge is all important. Sign it, I say, in the name of Heaven; but do not sign it because this, that, or the other temperate man has signed it, but because you feel it to be your only hope. Do it for yourself, and do it if you are the only man in the world who acts thus. To sign because another man, whom you think more respectable, has signed, will give you little or no strength. You must do it for yourself, and because it is right."

The parson was pretty ready with the tongue, and rarely came off second best when his opponents dragged him into a controversy, although his arguments were called by them, when he was not present, "mere fustian."

"His love for wine and brandy is at the bottom of all this hostility to the temperance cause," was boldly said of him by individuals in and out of his church. But especially were the members of other churches severe upon him.

"He'll turn out a drunkard," said one.

"I shouldn't be surprised to see him staggering in the streets before two years," said another.

"He does more harm to the temperance cause than ten drunkards," alleged a third.

While others said—"Isn't it scandalous!"

"He's a disgrace to his profession!"

"*He* pretend to have religion!"

"A minister indeed!"

And so the changes rang.

All this time, Mr. Manlius firmly maintained his ground, taking his glass of wine whenever it suited him. At last, after the occurrence of a dinner-party given by a family of some note in the place, at which the minister was present, and at which wine was circulated freely, a rather scandalous report got abroad, and soon went buzzing all over the village. A young man, who made no secret of being fond of his glass, and who was at the dinner-party, met, on the day after, a very warm advocate of temperance, and a member of a different denomination from that in which Mr. Manlius was a minister, and said to him, with mock gravity—"We had a *rara avis* at our dinner-party yesterday, Perkins."

"Indeed. What wonderful thing was that?"

"A tipsy parson."

"A what?"

The man's eyes became instantly almost as big as saucers.

"A tipsy parson."

"Who? Mr. Manlius?" was eagerly inquired.

"I didn't say so. I call no names."

"He was present, I know; and drank wine, I am told, like a fish."

"I wasn't aware before that fishes drank wine," said the man gravely.

"It was Manlius, wasn't it?" urged the other.

"I call no names," was repeated. "All I said was, that we had a tipsy parson—and so we had I'll prove it before a jury of a thousand, if necessary."

"It's no more than I expected," said the temperance man. "He's a mere winebibber at best. He pretend to preach the gospel! I wonder he isn't struck dead in the pulpit."

The moment his informant had left him, Perkins started forth to communicate the astounding intelligence that Mr. Manlius had been drunk on the day before, at Mr. Reeside's dinner-party. From lip to lip the scandal flew, with little less than electric quickness. It was all over the village by the next day. Some doubted, some denied, but the majority believed the story—it was so likely to be true.

This occurred near the close of the week, and Sunday arrived before the powers that be in the church were able to confer upon the subject, and cite the minister to appear and answer for himself on the scandalous charge of drunkenness. There was an unusual number of vacant pews during service, both morning and afternoon.

Monday came, and, early in the day, a committee of two deacons waited upon Mr. Manlius, and informed him of the report in circulation, and of their wish that he would appear before them on the next afternoon, to give an account of himself, as the church deemed the matter far too serious to be passed lightly over. The minister was evidently a good deal surprised and startled at this, but he neither denied the charge nor attempted any palliation, merely saying that he would attend, of course.

"It's plain that he's guilty," said Deacon Jones to Deacon Todd, as they walked with sober faces away from the minister's dwelling.

"Plain? Yes—it's written in his face," returned Deacon Todd. "So much for opposing temperance reforms and drinking wine. It's a judgment upon him."

"But what a scandal to our church!" said Deacon Jones.

"Yes—think of that. He must be suspended, and not restored until he signs the pledge."

"I don't believe he'll ever do that."

"Why not?"

"He says he would cut off his right hand first."

"People are very fond of cutting off their right hand, you know. My word for it, this will do the business for him. He will be glad enough to get the matter hushed up so easily. I shall go for suspending him until he signs the pledge."

"I don't know but that I will go with you. If he signs the pledge, he's safe."

And so the two deacons settled the matter.

On the next day, in grave council assembled were all the deacons of the church, besides sundry individuals who had come as the minister's friends or accusers. Perkins, who had put the report in circulation, was there, at the special request of one of the deacons, who had ascertained that he had as much, or a little more to say, in the matter, than any one.

Perkins was called upon, rather unexpectedly, to answer one or two questions, immediately on the opening of the meeting, but as he was a stanch temperance man, and cordially despised the minister, he was bold to reply.

"Mr. Perkins," said the presiding deacon, "as far as we can learn, this scandalous charge originated with you: I will, therefore, ask you—did you say that the Rev. Mr. Manlius was drunk at Mr. Reeside's dinner-party?"

"I did," was the unhesitating answer.

"Were you present at Mr. Reeside's?"

"No, sir."

"Did you see Mr. Manlius coming from the house intoxicated?"

"No."

"What evidence, then, have you of the truth of your charge? We have conversed this morning with several who were present, and all say that they observed nothing out of the way in Mr. Manlius, on the occasion of which you speak. This is a serious matter, and we should like to have your authority for a statement so injurious to the reputation of the minister and the cause of religion."

"My authority is Mr. Burton, who was present."

"Did he tell you that Mr. Manlius was intoxicated?"

"He said there was a drunken minister there, and Mr. Manlius, I have ascertained, was the only clergyman present."

"Was that so?" asked the deacon of an individual who was at Mr. Reeside's.

"Mr. Manlius was the only clergyman there," was replied.

"Then," said Perkins, "if there was a drunken minister there, it must have been Mr. Manlius. I can draw no other inference."

"Can Mr. Burton be found?" was now asked.

An individual immediately volunteered to go in search of him. In half an hour he was produced. As he entered the grave assembly, he looked around with great composure upon the array of solemn faces and eyes intently fixed upon him. He did not appear in the least abashed.

"You were at Mr. Reeside's last week, at a dinner-party, I believe?" said the presiding deacon.

"I was."

"Did you see Mr. Manlius intoxicated on that occasion?"

"Mr. Manlius! Good heavens! no! I can testify, upon oath, that he was as solemn as a judge. Who says that I made so scandalous an allegation?"

Burton appeared to grow strongly excited.

"I say so," cried Perkins in a loud voice.

"You say so? And, pray, upon what authority?"

"Upon the authority of your own words."

"Never!"

"But you did tell me so."

Perkins was much excited.

"When?"

"On the day after the dinner-party. Don't you remember what you said to me?"

"Oh, yes—perfectly."

"That you had a drunken minister at dinner?"

"No, I never said that."

"But you did, I can be qualified to it."

"I said we had a 'tipsy parson.'"

"And, pray, what is the difference?"

At the words "tipsy parson," the minister burst into a loud laugh, and so did two or three others who had been at Mr. Reeside's. The grave deacon in the chair looked around with frowning wonder at such indecorum, and felt that especially ill-timed was the levity of the minister.

"I do not understand this," he said, with great gravity.

"I can explain it," remarked an individual, rising, "as I happened to be at Mr. Reeside's, and know all about the 'tipsy parson.' The cook of our kind hostess, in her culinary ingenuity, furnished a dessert, which she called 'tipsy parson,'—made, I believe, by soaking sponge-cake in brandy and pouring a custard over it. It is therefore true, as our friend Burton has said, that there was a 'tipsy parson' at the table; but as to the drunken minister of Mr. Perkins, I know nothing."

Never before, in a grave and solemn assembly of deacons, was there such a sudden and universal burst of laughter, such a holding of sides and vibration of bodies, as followed this unexpected speech. In the midst of the confusion and noise, Perkins quietly retired. He has been known, ever since, in the village, much to his chagrin and scandalization, he being still a warm temperance man, as the "tipsy parson."

"There goes the 'tipsy parson,'" he hears said, as he passes along the street, a dozen times in a week, and he is now seriously inclined to leave the village, in order to escape the ridicule his over-zealous effort to blast the minister's reputation has called into existence. As for the Rev. Mr. Manlius, he often tells the story, and laughs over it as heartily as any one.

MUCH ADO ABOUT NOTHING;

OR, THE REASON WHY MRS. TODD DIDN'T SPEAK TO MRS. JONES.

"DID you see that?" said Mrs. Jones to her friend Mrs. Lyon, with whom she was walking.

"See what?"

"Why, that Mrs. Todd didn't speak to me."

"No. I thought she spoke to you as well as to me."

"Indeed, then, and she didn't."

"Are you sure?"

"Sure? Can't I believe my own eyes? She nodded and spoke to you, but she didn't as much as look at me."

"What in the world can be the reason, Mrs Jones?"

"Dear knows!"

"You certainly must be mistaken. Mrs. Todd would not refuse to speak to one of her old friends in the street."

"Humph! I don't know; she's rather queer, sometimes. She's taken a miff at something, I suppose, and means to cut my acquaintance. But let her. I shall not distress myself about it; she isn't all the world."

"Have you done any thing likely to offend her?" asked Mrs. Lyon.

"Me?" returned her companion. "No, not that I am aware of; but certain people are always on the lookout for something or other wrong, and Mrs. Todd is just one of that kind."

"I never thought so, Mrs. Jones."

"She is, then. I know her very well."

"I'm sorry," said Mrs. Lyon, evincing a good deal of concern. "Hadn't you better go to her in a plain, straight-forward way, and ask the reason of her conduct? This would make all clear in a moment."

"Go to her, Mrs. Lyon," exclaimed Mrs. Jones, with ill-concealed indignation. "No, indeed, that I will not. Do you think I would demean myself so much?"

"I am not sure that by so doing you would demean yourself, as you say. There is, clearly, some mistake, and such a course would correct all false impressions. But it was only a suggestion, thrown out for your consideration."

"Oh, no, Mrs. Lyon," replied Mrs. Jones, with warmth. "You never find me cringing to people, and begging to know why they are pleased to cut my acquaintance. I feel quite as good as anybody, and consider myself of just as much consequence as the proudest and best. Mrs. Todd needn't think I care for her acquaintance; I never valued it a pin."

Notwithstanding Mrs. Jones's perfect indifference toward Mrs. Todd, she continued to talk about her, pretty much after this fashion, growing more excited all the while, during the next half hour, at the close of which time the ladies parted company.

When Mrs. Jones met her husband at the dinner-table, she related what had happened during the morning. Mr. Jones was disposed to treat the matter lightly, but his wife soon satisfied him that the thing was no joke.

"What can be Mrs. Todd's reason for such conduct?" he asked, with a serious air.

"I can't tell, for my life."

"She must have heard some false report about you."

"It's as likely as not; but what can it be?"

"Something serious, to cause her to take so decided a stand as she seems to have done."

Mr. Jones looked grave, and spoke in a grave tone of voice. This made matters worse. Mrs. Jones's first idea was that Mrs. Todd had heard something that she might have said about her, and that wounded pride had caused her to do as she had done; but her husband's remark suggested other thoughts. It was possible that reports were in circulation calculated to injure her social standing, and that Mrs. Todd's conduct toward her was not the result of any private pique.

"It is certainly strange and unaccountable," she said, in reply to her husband's last remark, speaking in a thoughtful tone.

"Would it not be the fairest and best way for you to go and ask for an explanation?"

"No, I can't do that," replied Mrs. Jones, quickly. "I am willing to bear undeserved contempt and unjust censure, but I will never humble myself to any one."

For the rest of the day, Mrs. Jones's thoughts all flowed in one channel. A hundred reasons for Mrs. Todd's strange conduct were imagined, but none seemed long satisfactory. At last, she remembered having spoken pretty freely about the lady to a certain individual who was not remarkable for his discretion

"That's it," she said, rising from her chair, and walking nervously across the floor of her chamber, backward and forward, for two or three times, while a burning glow suffused her cheek. "Isn't it too bad that words spoken in confidence should have been repeated! I don't wonder she is offended."

This idea was retained for a time, and then abandoned for some other that seemed more plausible. For the next two weeks, Mrs. Jones was very un happy. She did not meet Mrs. Todd during that period, but she saw a number of her friends, to whom either she or Mrs. Lyon had communicated the fact already stated. All declared the conduct of Mrs. Todd to be unaccountable; but several, among themselves, had shrewd suspicions of the real cause. Conversations on the subject, like the following, were held:—

"I can tell you what I think about it, Mrs. S——. You know, Mrs. Jones is pretty free with her tongue?"

"Yes."

"You've heard her talk about Mrs. Todd?"

"I don't remember, now."

"I have, often; she doesn't spare her, sometimes. You know, yourself, that Mrs. Todd has queer ways of her own."

"She is not perfect, certainly."

"Not by a great deal; and Mrs. Jones has not hesitated to say so. There is not the least doubt in my mind, that Mrs. Todd has heard something."

"Perhaps so; but she is very foolish to take any notice of it."

"So I think; but you know she is touchy."

In some instances, the conversation assumed a grave form:—

"Do you know what has struck me, in this matter of Mrs. Jones and Mrs. Todd?" says one scandal-loving personage to another, whose taste ran parallel with her own.

"No. What is it?" eagerly asks the auditor.

"I will tell you; but you mustn't speak of it, for your life."

"Never fear me."

The communication was made in a deep whisper.

"Bless me!" exclaims the recipient of the secret. "It surely cannot be so!"

"There is not the least doubt of it. I had it from a source that cannot be doubted."

"How in the world did you hear it?"

"In a way not dreamed of by Mrs. Jones."

"No doubt, Mrs. Todd has heard the same."

"Not the least in the world. But don't you think her to blame in refusing to keep Mrs. Jones's company, or even to speak to her?"

"Certainly I do. It happened a long time ago, and no doubt poor Mrs. Jones has suffered enough on account of it. Indeed, I don't think *she* ought to be blamed in the matter at all; it was her misfortune, not her fault."

"So I think. In fact, I believe she is just as worthy of respect and kindness as Mrs. Todd."

"No doubt of it in the world; and from me she shall always receive it."

"And from me also."

In this way the circle spread, so that before two weeks had elapsed, there were no less than twenty different notions held about Mrs. Todd's behaviour to Mrs. Jones. Some talked very seriously about cutting the acquaintance of Mrs. Jones also, while others took her side and threatened to give up the acquaintance of Mrs. Todd.

Thus matters stood, when a mutual friend, who wished to do honour to some visitors from a neighbouring city, sent out invitations for a party. Before these invitations were despatched, it was seriously debated whether it would do to invite both Mrs. Jones and Mrs. Todd, considering how matters stood between them. The decision was in favour of letting them take care of their own difficulties.

"If I thought Mrs. Todd would be there, I am sure I wouldn't go," said Mrs. Jones, on receiving her card of invitation.

"I hardly think that would be acting wisely," replied her husband. "You are not conscious of having wronged Mrs. Todd. Why, then, should you shun her?"

"But it is so unpleasant to meet a person with whom you have been long intimate, who refuses to speak to you."

"No doubt it is. Still we ought not to go out

of our way to shun that person. Let us, while we do not attempt to interfere with the liberties of others, be free ourselves. Were I in your place, I would not move an inch to keep out of her way."

"I have not your firmness. I wish I had. It was only yesterday that I crossed the street to keep from meeting her face to face."

"You were wrong."

"I can't help it; it is my weakness. Three times already have I put myself about to avoid her; and if I could frame any good excuse for staying away from this party, I certainly should do so. I would give any thing for a good sick-headache on Tuesday next."

"I am really ashamed of you, Ellen. I thought you more of a woman," said Mr. Jones.

The night of the party at length came round. During the whole day preceding it, Mrs. Jones could think of nothing but the unpleasant feelings she would have upon meeting with Mrs. Todd, and her "heart was in her mouth" all the time. She wished a dozen times that it would rain. But her wishes availed nothing; not a cloud was to be seen in the clear blue firmament from morning until evening.

"Oh, if I only had some good excuse for staying at home!" she said over and over again; but no good excuse offered.

Mr Jones saw that his wife was in a very unhappy state of mind, and tried his best to cheer her, but with little good effect.

"It is no use to talk to me, I can't help it," she replied to his remonstrance, in a husky voice. "I am neither a stock nor a stone."

"There's Mrs. Jones," said one friend to another, on seeing the lady they named enter Mrs. ——'s well-filled parlours.

"Where is Mrs. Todd?" asked the lady addressed.

"Sure enough! where is she?" replied the other. "Oh, there she is, in the other room. I wonder why it is that she does not speak to Mrs. Jones."

"No one knows."

"It's very strange."

"I'll tell you what I've heard."

"What?"

"That she's jealous of Mrs. Jones."

"Ridiculous!"

"Isn't it."

"I don't believe a word of it."

"Nor I. I only told you what I had heard."

"There must be some other reason."

"And doubtless is."

Meantime, Mrs. Jones found a seat in a corner, where she ensconced herself, with the determination of keeping her place during the evening, that she might avoid the unpleasantness of coming in contact with Mrs. Todd. All this was, of course, very weak in Mrs. Jones. But she had no independent strength of character, it must be owned.

"Poor Mrs. Jones! How cut down she looks," remarked a lady who knew all about the trouble that existed. "I really feel sorry for her."

"She takes it a great deal too much to heart," was the reply. "Mrs. Todd might refuse to speak to me a dozen times, if she liked. It wouldn't break my heart. But where is she?"

"In the other room, as gay and lively as ever I saw her. See, there she is."

"Yes, I see her. Hark! You can hear her laugh to here. I must confess I don't like it. I don't believe she has any heart. She must know that Mrs. Jones is hurt at what she has done."

"Of course she does, and her manner is meant to insult her."

Seeing the disturbed and depressed state of Mrs. Jones's mind, two or three of her friends held a consultation on the subject, and finally agreed that they would ask Mrs. Todd, who seemed purposely to avoid Mrs. Jones, why she acted towards her as she did. But before they could find an opportunity of so doing, a messenger came to say that one of Mrs. Todd's children had been taken suddenly ill. The lady withdrew immediately.

Mrs. Jones breathed more freely on learning that Mrs. Todd had gone home. Soon after, she emerged from her place in the corner, and min gled with the company during the rest of the evening.

Mrs. Todd, on arriving at home, found one of her children quite sick; but it proved to be nothing serious. On the following morning, the little fellow was quite well again.

On that same morning, three ladies, personal friends of Mrs. Todd, met by appointment, and entered into grave consultation. They had undertaken to find out the cause of offence that had occurred, of so serious a character as to lead Mrs. Todd to adopt so rigid a course towards Mrs. Jones, and, if possible, to reconcile matters.

"The sickness of her child will be a good excuse for us to call upon her," said one. "If he is better, we can introduce the matter judiciously."

"I wonder how she will take it?" suggested another.

"Kindly, I hope," remarked the third.

"Suppose she does not?"

"We have done our duty."

"True. And that consciousness ought to be enough for us."

"She is a very proud woman, and my fear is, that, having taken an open and decided stand, she will yield to neither argument nor persuasion. Last night she over-acted her part. While she carefully avoided coming in contact with Mrs. Jones, she was often near her, and on such occasions talked and laughed louder than at any other time. I thought once or twice, that there was something of malice exhibited in her conduct."

To this, one of the three assented. But the other thought differently. After some further discussion, and an ineffectual attempt to decide which of them should open the matter to Mrs. Todd, the ladies sallied forth on their errand of peace. They found Mrs. Todd at home, who received them in her usual agreeable manner.

"How is your little boy?" was the first question, after the first salutations were over.

"Much better than he was last night, I thank you. Indeed, he is quite as well as usual."

"What was the matter with him, Mrs. Todd?"

"It is hard to tell. I found him with a high fever, when I got home. But it subsided in the course of an hour. Children often have such attacks. They will be quite sick one hour, and apparently well the next."

"I am very glad to hear that it is nothing serious," said one of the ladies. "I was afraid it might have been croup, or something as bad."

There was a pause.

"It seemed a little unfortunate," remarked one of the visitors, "for it deprived you of an evening's enjoyment."

"Yes, it does appear so, but no doubt it is all right. I suppose you had a very pleasant time?"

"Oh, yes. Delightful!"

"I hadn't seen half my friends when I was summoned away. Was Mrs. Williams there?"

"Oh, yes."

"And Mrs. Gray?"

"Yes."

"And Mrs. Elder?"

"Yes."

"I didn't see either of them."

"Not a word about Mrs. Jones," thought the ladies

A light running conversation, something after this style, was kept up, with occasional pauses, for half an hour, when one of the visitors determined to come to the point.

"Mrs. Todd—a-hem!" she said in one of the pauses that always take place in uninteresting conversation.

The lady's tone of voice had so changed from what it was a few moments before, that Mrs. Todd looked up at her with surprise. No less changed was the lady's countenance. Mrs. Todd was mistified. But she was not long in doubt.

"A-hem! Mrs. Todd, we have come to—to—as friends—mutual friends—to ask you"—

The lady's voice broke down; but two or three "a-hems!" partially restored it, and she went on.

"To ask why you refused to—to—speak to Mrs. Jones?"

"Why I refused to speak to Mrs. Jones?" said Mrs. Todd, her cheek flushing.

"Yes. Mrs. Jones is very much hurt about it, and says she cannot imagine the reason. It has made her very unhappy. As mutual friends, we have thought it our duty to try and reconcile matters. It is on this errand that we have called this morning. Mrs. Jones says she met you for the last time about two weeks ago, and that you refused to speak to her. May we ask the reason."

"You may, certainly," was calmly replied.

Expectation was now on tiptoe.

"What, then, was the reason?"

"*I did not see her.*"

"What? Didn't you refuse to speak to her?"

"Never in my life. I esteem Mrs. Jones too highly. If I passed her, as you say, without speaking, it was because I did not see her."

In less than half an hour, Mrs. Todd was at the house of Mrs. Jones. What passed between the ladies need not be told.

ALMOST A TRAGEDY.

A REMINISCENCE OF MR. JOHN JONES.

It is now about five years since I met with a little adventure in the West, which may be worth relating. It caused me a good deal of excitement at first; regrets afterward, for the temporary pain I inflicted, and many a hearty laugh since. New things come up so rapidly that it is almost impossible to keep the run of them, and it is not at all surprising that those who are content to go along in the good old way should now and then be caught napping. I own that I was, completely.

Business took me out West, in the spring of 18—, and kept me in Ohio for the entire summer of that year. After a hard day's ride, in the month of August, I entered, just before nightfall, a certain town lying on the National Road, where I expected to remain for a week. After taking possession of my room at the hotel; shaving, washing, and improving my appearance in other respects, I came down and took a seat in the porch that ran along the front of the house. I had not been here very long before the stage from the East drove up, and the passengers, who were to take supper, as this was a stage-house, alighted. Among them, I noticed a woman with a pale, emaciated, and, I would have said, dying child in her arms. Her face was anxious and haggard in its expression. She was accompanied by a man, whom I rightly supposed to be her husband. He immediately went to the bar and engaged a room, saying that his child was too sick to permit them to continue their journey.

"Do you wish a doctor?" asked the landlord.

"No," replied the man. "We have medicine, prescribed by our own physician before we left home. If that does no good, we have little confidence in any other remedies."

No more was said. The man was shown to his room, whither he retired with his wife and sick child. The room, it so happened, was next to mine, and the two rooms communicated by a door, which was of course closed and fastened.

The emaciated child and anxious mother presented a sight that fixed itself upon my mind, and excited my liveliest sympathies. I could not get them from my thoughts.

About ten o'clock that night, I took a candle and went to my room. Before undressing myself, I sat down at a table to make some entries of collections and expenses, and to think over and arrange my business for the next day. All was still, except now and then a slight movement in the next chamber, where the parents were sitting up with their sick child.

"What did you give him last?" I heard the father say, in a low, but distinct tone

"Aconite," was as distinctly replied

This I knew to be a deadly poison. I listened, you may be sure, with a more earnest attention

"How many grains?" was next asked

"Two," replied the mother.

Two grains of aconite! My hair began to rise.

"I think we had better increase the dose to five grains."

Horrible!

"It's an hour since he took the last, and I see no change," said the mother. "Perhaps we had better try the arsenic."

My blood ran cold at this murderous proposition I felt like starting up, bursting open the door, and confronting them in their dreadful work. But, as if spell-bound, I remained where I was. To the last proposition, the man replied—

"I would rather see the aconite tried in a larger dose. If, in half an hour, there is no visible effect from it, then we will resort to the arsenic."

"If you think it best," said the mother, in a low sad voice—(well she might be sad over such awful work)—"let us try the aconite again, but in a larger dose. You will find it on the mantelpiece."

I heard the deliberate tread of the man, as he crossed the room for a larger dose of the poison, while I hurriedly deliberated the question of what I should do. Before I could make up my mind to act, I heard his returning step. A few moments of awful stillness succeeded. I felt as if I were in the centre of a sphere, with the gravitating forces from every point of the circumference upon me. I don't think I could have moved a limb to save my life.

"There; let us see what they will do," came distinctly upon my ear.

Gracious Heaven! the deed was done. Five grains of aconite given to the tender child, already on the verge of death! The cold sweat came out over my whole body, and stood in clammy drops upon my forehead. All was still. Death was doing his awful work in silence. I sat motionless, under the influence of a strange irresolution or imbecility of mind, unable to determine what steps to take in a matter where all now seems as plain to me as daylight. I do not know what came over me. The fact only shows how, when placed in certain positions, we become paralyzed, and unable to act even with common decision. I remember saying to myself, as a justification for not interfering at this stage of the proceedings—

"It is too late now. Five and three are eight

Eight grains of aconite! There is no longer a vestige of hope for the child. Death is as certain as if a bullet were fired through the sufferer's head."

I did not stir from where I sat, but tried to hush my deep breathing, and quiet the loud pulsations of my heart, lest even they should be heard and betray my proximity to the wretches.

Half an hour passed. There was a movement, and the murmuring sound of voices,—but, though I listened eagerly, I was not able to make out what was said. I heard the tread of a man across the floor, and I also heard his return. I thought of the arsenic, and said to myself, at the same time, "They will not need that." The woman was speaking. I listened.

"Was that the arsenic?"

"Yes."

"How many grains did you give him?"

"I meant to give him three, but, in mistake, gave him six or seven."

It was too late, now, for any interference. But, I was determined that the wretches should not escape. I was an ear-witness to their murderous act, and I resolved to bring them to the light. While I thus mused and resolved, I was thrilled by a long, tremulous cry from the dying child. All was again still as death, save an occasional deep sob, that seemed bursting up from the remnant of stifled nature in the mother's bosom. Again that cry arose suddenly on the air, but feebler and shorter. The mother's sob now became a moan, and soon changed to a low, wailing cry. Her child was dead. The fatal drugs had too surely done their murderous work. But why should she weep over the precious babe her own hand had destroyed? and why came there, now and then, from that chamber of death, a deep sighing moan, struggling up in spite of all efforts to repress it, from the breast of the miserable father? Strange enigma! I could not read, satisfactorily to myself, the difficult solution.

I still remained quiet where I was. In a little while I heard the father go out, and listened to his footsteps until they became lost in silence. Soon the hasty tread of several feet were heard, and two or three females entered the rooom. Their presence caused the woman to cry bitterly.

"False-hearted, cruel wretch!" I could not help muttering to myself. "Hypocritical cries and crocodile tears will not hide your sin. An ear of which you dreamed not has heard your hellish plots, and been witness to your hellish deeds upon the body of your poor babe. You cannot escape. The voice of blood cries from the very ground. The hope of the murderer is vain. He cannot hide himself from the pursuer."

For half the night, I lay awake, thinking of what had occurred, and settling in my mind the course of proceeding to adopt in the morning. I was up long before sunrise—in fact, long before anybody else was stirring—awaiting the appearance of the landlord, to whom it was my intention to give information of the dreadful deed that had been committed. Full an hour elapsed before he made his appearance. I immediately drew him aside.

"There has been a death in the house," said I.

"Yes," he replied. "The poor sick child that was brought here by the Eastern stage last evening died in the night. I did not suppose it would live till morning. To me, it seemed in a dying state when its parents arrived."

"There has been foul play," said I, with emphasis. "That child has not died a natural death."

"How so? What do you mean?" asked the landlord, with a look of surprise.

"I mean what I say," was my reply. "As sure as I am a living man, that child has been murdered."

I then related all I had heard, to the horror and astonishment of the landlord.

"A deed like this must not go unpunished," he said, sternly and angrily. "It is horrible to think of it."

After talking over the matter for some time, it was determined to call a council of half a dozen of the regular boarders in the house, as soon as breakfast was over, and decide upon the steps best to be taken. Accordingly, after breakfast, a few of us assembled in a private parlour, and I again related, with minuteness, all that I had heard. After sundry expressions of horror and indignation, a gentleman said to me—"Are you sure it was grains or granules of aconite and arsenic that were given to the child?"

"Grains, sir," I replied, promptly.

"This is a serious matter," he added; "and if there should be any mistake, it would be sad indeed to harrow the feelings of those bereaved parents by so dreadful a charge as that of the murder of their own offspring. My own impression is, that our friend here is under a mistake."

"Can't I believe my own ears, sir?" said I, a little indignantly.

"Don't misunderstand me," returned the gentleman, politely. "I don't doubt you have heard all you say, and it may be even to the word grains; but I am under the impression that the arsenic and aconite given were in the homœopathic preparations, and therefore no longer poisonous."

There was a long pause after this was said; every one present seemed to breathe more freely. I had heard of homœopathy, and something about infinitesimal doses, but had never seen the medicine used, neither did I know any thing about the mode in which it was sometimes practised.

"Suppose we send for the man," suggested the landlord, "and question him, but in a way not to wound him, if he be innocent."

This, after some debate, was agreed upon, and a servant was sent to his room with a request that he would come to the parlour. He obeyed the sum mons instantly, but looked a good deal surprised when he saw a grave assembly of six or seven persons. The gentleman who had expressed the doubt in the man's favour, said to him, as soon as he had taken his seat—"We have learned, sir, with sincere regret, that you were so unfortunate as to lose your child last night—a severe affliction. Though strangers, we deeply sympathize with you."

The man expressed his thanks, in a few words, for the kind feelings manifested, and said that, as it was their only child, they felt the affliction more severely, but were still willing to submit to the loss, as a Divine dispensation, grievous to be borne, yet intended for good.

"You did not call in a physician," said the individual who had at first addressed him.

"No," replied the man. "Before starting for Cincinnati, yesterday morning, we learned that, no matter how ill our child might become, we could not get the advice of a homœopathic physician until we reached home, and we were not willing to trust our child in the hands of any other. We, therefore, before commencing our journey, obtained medicine, and advice how to administer it should alarming symptoms occur."

"Homœopathic medicines?"

"Yes, sir."

"In powders, I suppose?"

"No, sir; in little grains or pellets, like these."

And he drew from his pocket a diminutive vial, the smallest I had ever seen, in which were a number of little white granules, about the size of the head of a pin. A printed label was wound around the vial, and it bore the word "Arsenicum." It passed from hand to hand, and all read it.

"You gave this?" said the volunteer spokesman.

"Yes, sir; that and aconite."

"How much is a dose?"

"From one to five or six grains."

"Or granules?"

"Yes."

The little bottle was returned to the man, who placed it in his pocket. A pause ensued. The truth was plain enough to us all. The individual whose sagacity, or better information about what was going on in the world, had saved a most painful denouement to this affair, said to the man, in a way as little as possible calculated to wound his feelings—

"You are, of course, surprised at this proceeding—this seemingly wanton intrusion upon your grief. But you will understand it when I tell you, that a lodger, in a room adjoining yours, who knew nothing of homœopathy, heard you speak of giving your child several grains of aconite and arsenic You can easily infer the impression upon his mind This morning, he related what he had heard, when an individual here present, who suspected the truth, suggested that you be sent for and asked the questions which you have so satisfactorily answered. Do not, let me beg of you, feel hurt. What we have done was but an act of justice to yourself."

The man smiled sadly, and, thanking us with eyes fast filling with tears, rose up quickly to conceal his emotion, and retired from the room.

"Landlord," said I, an hour afterwards, "I want my valise taken out of No. 10, and put into some other room."

"Why so? Isn't the room a pleasant one?"

"Oh, yes; but I'd like a change."

"Very well; we'll put you in No. 16."

I was the "lodger in the room adjoining," and didn't, therefore, wish to appear on the premises and be known by the man, as the getter up of a suspicion against him. I did not come home to dinner, and kept out of the way till after dark. When I returned to the hotel, I was relieved to find that the bereaved parents had departed with the dead body of their child. But the whole company were now at liberty to laugh at what had occurred to their hearts' content, and to laugh at me in particular. I stood it that evening, as well as I could; but finding, on the next day, that it was renewed with as keen a zest as ever, concluded to close up my business on the spot, and leave the place—which I did.

THAT JOHN MASON.

"WHAT kind of people have you here?" I asked of one of my first acquaintances, after becoming a denizen of the pleasant little village of Moorfield.

"Very clever people, with one or two exceptions," he replied. "I am sure you will like us very well."

"Who are the exceptions?" I asked. "For I wish to keep all such exceptions at a distance. Being a stranger, I will, wisely, take a hint in time. It's an easy matter to shun an acquaintanceship; but by no means so easy to break it off, after it is once formed."

"Very truly said, Mr. Jones. And I will warn you, in time, of one man in particular. His name is John Mason. Keep clear of him, if you wish to keep out of trouble. He's as smooth and oily as a whetstone; and, like a whetstone, abrades every thing he touches. He's a bad man, that John Mason."

"Who, or what is he?" I asked.

"He's a lawyer, and one of the principal holders

of property in the township. But money can't gild him over. He's a bad man, that John Mason, and my advice to you and to every one, is to keep clear of him. I know him like a book."

"I'm very much obliged to you," said I, "for your timely caution: I will take care to profit by it."

My next acquaintance bore pretty much the same testimony, and so did the next. It was plain that John Mason was not the right kind of a man, and rather a blemish upon the village of Moorfield, notwithstanding he was one of the principal property-holders in the township.

"If it wasn't for that John Mason," I heard on this hand, and, "If it wasn't for that John Mason!" I heard on the other hand, as my acquaintanceship among the people extended. Particularly bitter against him was the first individual who had whispered in my ear a friendly caution; and I hardly ever met with him, that he hadn't something to say about that John Mason.

About six months after my arrival in Moorfield, I attended a public meeting, at which the leading men of the township were present. Most of them were strangers to me. At this meeting, I fell in company with a very pleasant man, who had several times addressed those present, and always in such a clear, forcible, and common-sense way, as to carry conviction to all but a few, who carped and quibbled at every thing he said, and in a very churlish manner. Several of those quibblers I happened to know. He represented one set of views, and they another. His had regard for the public good; theirs looked, it was plain, to sectional and private interests.

"How do you like our little town, Mr. Jones?" said this individual to me, after the meeting had adjourned, and little knots of individuals were formed here and there for conversation.

"Very well," I replied.

"And the people?" he added.

"The people," I answered, "appear to be about a fair sample of what are to be found everywhere. Good and bad mixed up together."

"Yes. That, I suppose, is a fair general estimate."

"Of course," I added, "we find, in all communities, certain individuals, who stand out more prominent than the rest—distinguished for good or evil. This appears to be the case here, as well as elsewhere."

"You have already discovered, then, that, even in Moorfield, there are some bad men."

"Oh, yes. There's that John Mason, for instance."

The man looked a little surprised, but remarked, without any change of tone—"So, you have heard of him, have you?"

"Oh, yes."

"As a very bad man?"

"Of course. You know him, I suppose?"

"Yes, very well. Have you ever met him?"

"No, and never wish to."

"You've seen him, I presume?"

"Never. Is he here?"

The man glanced round the room, and then re plied—"I don't see him."

"He was here, I suppose?"

"Oh, yes, and addressed the meeting several times."

"In one of those sneering, ill-tempered answers to your remarks, no doubt."

The man slightly inclined his head, as if acknowledging a compliment.

"It's a pity," said I, "that such men as this John Mason often have wealth and some shrewdness of mind to give them power in the community."

"Perhaps," said my auditor, "your prejudices against this man are too strong. He's not perfect, I know; but even the devil is often painted blacker than he is. If you knew him, I rather think you would estimate him a little differently."

"I don't wish to know him. Opportunities have offered, but I have always avoided an introduction."

"Who first gave you the character of this man?" asked the individual with whom I was conversing.

"Mr. Laxton," I replied. "Do you know him?"

"Oh, yes, very well. He speaks hard of Mason, does he?"

"He has cause, I believe."

"Did he ever explain to you what it was?"

"Not very fully; but he gives him a general bad character, and says he has done more to injure the best interests of the village than any ten of its worst enemies that exist."

"Indeed! That is a sweeping declaration. But I will frankly own that I cannot join in so broad a condemnation of the man, although he has his faults, and no one knows them, I think, better than I do."

This made no impression upon me. The name of John Mason was associated in my mind with every thing that was bad, and I replied by saying that I was very well satisfied in regard to his character, and didn't mean to have any thing to do with him while I lived in Moorfield.

Some one interrupted our conversation at this point, and I was separated from my very agreeable companion. I met him frequently afterwards, and he was always particularly polite to me, and once or twice asked me if I had fallen in with that John Mason yet; to which I always replied in the negative, and expressed myself as ever in regard to the personage mentioned.

Careful as we may be to keep out of trouble, we are not always successful in our efforts. When I removed to Moorfield, I supposed my affairs to be in a very good way; but things proved to be otherwise. I was disappointed, not only in the amount

I expected to receive from the business I followed in the village, but disappointed in the receipt of money I felt sure of getting by a certain time.

When I first came to Moorfield, I bought a piece of property from Laxton—this business transaction made us acquainted—and paid, cash down, one-third of the purchase-money, the property remaining as security for the two-thirds, which I was under contract to settle at a certain time. My first payment was two thousand dollars. Unfortunately, when the final payment became due, I was not in funds, and the prospect of receiving money within five or six months was any thing but good. In this dilemma, I waited upon Laxton, and informed him of my disappointment. His face became grave.

"I hope it will not put you to any serious inconvenience."

"What?" he asked.

"My failure to meet this payment on the property. You are fully secured, and within six months I will be able to do what I had hoped to do at this time."

"I am sorry, Mr. Jones," he returned, "but I have made all my calculations to receive the sum due at this time, and cannot do without it."

"But I haven't the money, Mr. Laxton, and have fully explained to you the reason why."

"That is your affair, not mine, Mr. Jones. If you have been disappointed at one point, it is your business to look to another. A contract is a con tract."

"Will you not extend the time of payment?" said I.

"No, sir, I cannot."

"What will you do?"

"Do? You ask a strange question."

"Well, what will you do?"

"Why, raise the money on the property."

"How will you do that?"

"Sell it, of course."

I asked no further questions, but left him and went away. Before reaching home, to which place I was retiring in order to think over the position in which I was placed, and determine what steps to take, if any were left to me, I met the pleasant acquaintance I had made at the town-meeting.

"You look grave, Mr. Jones," said he, as we paused, facing each other. "What's the matter?"

I frankly told him my difficulty.

"So Laxton has got you in his clutches, has he?" was the simple, yet, I perceived, meaning reply that he made.

"I am in his clutches, certainly," said I.

"And will not get out of them very easily, I apprehend."

"What will he do?"

"He will sell the property at auction."

"It won't bring his claim under the hammer."

"No, I suppose not, for that is really more than the property is worth."

"Do you think so?"

"Certainly I do. I know the value of every lot of ground in the township, and know that you have been taken in in your purchase."

"What do you suppose it will bring at a forced sale?"

"Few men will bid over twenty-five hundred dollars.

"You cannot be serious?"

"I assure you I am. He, however, will overbid all, up to four thousand. He will, probably, have it knocked down to him at three thousand, and thus come into the unencumbered possession of a piece of property upon which he has received two thousand dollars."

"But three thousand dollars will not satisfy his claim against me."

"No. You will still owe him a thousand dollars."

"Will he prosecute his claim?"

"He?" And the man smiled. "Yes, to the last extremity, if there be hope of getting any thing."

"Then I am certainly in a bad way."

"I'm afraid you are, unless you can find some one here who will befriend you in the matter."

"There is no one here who will lend me four thousand dollars upon that piece of property," said I.

"I don't know but one man who is likely to do it," was answered.

"Who is that?" I asked, eagerly.

"John Mason."

"John Mason! I'll never go to him."

"Why not?"

"I might as well remain where I am as get into his hands—a sharper and a lawyer to boot. No, no. Better to bear the evils that we have, than fly to others that we know not of."

"You may get assistance somewhere else, but I am doubtful," said the man; and, bowing politely, passed on, and left me to my own unpleasant reflections.

Laxton made as quick work of the business as the nature of the case would admit, and in a very short time the property was advertised at public sale. As the time for the sale approached, the great desire to prevent the sacrifice that I was too well assured would take place, suggested the dernier resort of calling upon Mason; but my prejudice against the man was so strong, that I could not get my own consent to do so.

On the day before the sale, I met the individual before alluded to.

"Have you been to see Mason?" he asked

I shook my head.

"Then you have made up your mind to let that scoundrel, Laxton, fleece you out of your property?"

"I see no way of preventing it."

"Why don't you try Mason?"

"I don t believe it would do any good."

"I think differently."

"If he did help me out of this difficulty," I replied, "it would only be to get me into a more narrow corner."

"You don't know any such thing," said the man, in a different tone from any in which he had yet spoken when Mason was the subject of our remarks "Think, for a moment, upon the basis of your pie-judice; it lies mainly upon the assertion of Laxton, whom your own experience has proved to be a scoundrel. The fact is, your estimate of Mason's character is entirely erroneous. Laxton hates him, be cause he has circumvented him more than a dozen times in his schemes of iniquity, and will circumvent him again, if I do not greatly err, provided you give him the opportunity of doing so."

There was force in the view. True enough; what confidence was there to be placed in Laxton's words? And if Mason had circumvented him, as was alleged, of course there was a very good reason for detraction.

"At what hour do you think I can see him?" said I.

"I believe he is usually in about twelve o'clock."

"I will see him," said I, with emphasis.

"Do so," returned the man; "and may your interview be as satisfactory as you can desire."

At twelve, precisely, I called upon Mason, not without many misgivings, I must own. I found my prejudices still strong; and as to the good result, I could not help feeling serious doubts. On entering his office, I found no one present but the individual under whose advice I had called.

"Mr. Mason is not in," said I, feeling a little disappointed.

"Oh, yes, he is in," was replied.

I looked around, and then turned my eyes upon the man's face. I did not exactly comprehend its expression.

"My name is John Mason," said he, bowing politely; "so be seated, and let us talk over the business upon which you have called on me."

I needed no invitation to sit down, for I could not have kept my feet if I had tried, so suddenly and completely did his words astonish and confound me.

I will not repeat the confused, blundering apologies I attempted to make, nor give his gentlemanly replies. Enough, that an hour before the time at which the sale was advertised to take place on the next day, I waited upon Laxton.

"Be kind enough," said I, "to let me have that obligation upon which your present stringent measures are founded. I wish to take it up."

The man looked perfectly blank.

"Mr. John Mason," said I, "has generously furnished me with the funds necessary to save my property from sacrifice, and will take the securities you hold."

"Blast that John Mason!" ejaculated Laxton, with excessive bitterness, turning away and leaving me where I stood. I waited for ten minutes, but he did not come back. A suspicion that he meant to let the sale go on, if possible, crossed my mind. and I returned to Mason, who saw the sheriff and had the whole matter arranged.

Laxton has never spoken to me since. As for "That John Mason," I have proved him to be a fast friend, and a man of strict honour in every thing. So much for slander.

A NEW WAY TO COLLECT AN OLD DEBT.

EARLY in life, Mr. Jenkins had been what is called unfortunate in business. Either from the want of right management, or from causes that he could not well control. he became involved, and was broken all to pieces. It was not enough that he gave up every dollar he possessed in the world. In the hope that friends would interfere to prevent his being sent to jail, some of his creditors pressed eagerly for the balance of their claims, and the unhappy debtor had no alternative but to avail himself of the statute made and provided for the benefit of individuals in his extremity. It was a sore trial for him; but any thing rather than to be thrown into prison.

After this tempest of trouble and excitement, there fell upon the spirits of Mr. Jenkins a great calm. He withdrew himself from public observation for a time, but his active mind would not let him remain long in obscurity. In a few months, he was again in business, though in a small way. His efforts were more cautiously directed than before, and proved successful. He made something above his expenses during the first year, and after that accumulated money rapidly. In five or six years, Mr. Jenkins was worth some nine or ten thousand dollars.

But with this prosperity came no disposition on the part of Mr. Jenkins to pay off his old obligations. "They used the law against me," he would say, when the subject pressed itself upon his mind, as it would sometimes do, "and now let them get what the law will give them."

There was a curious provision in the law by which Jenkins had been freed from all the claims of his creditors against him; and this provision is usually incorporated in all similar laws, though for what reason it is hard to tell. It is only necessary to promise to pay a claim thus annulled, to bring it in full force against the debtor. If a man owes another a hundred dollars, and, by economy and self-denial, succeeds in saving twenty dollars and paying them to him, he becomes at once liable for the remaining eighty dollars, unless the manner of doing it be very guarded, and is in danger of a prosecution, although unable to pay another cent. A prudent man, who has once been forced into the unhappy alternative of taking the benefit of the insolvent law, is always careful, lest, in an unguarded moment, he acknowledge his liability to some old creditor, before he is fully able to meet it Anxious as he is to assure this one and that one of his desire and intention to pay them, if ever in his power, and to say to them that he is struggling early and late for their sakes as well as his own, his lips must remain sealed. A word of his intentions, and all his fond hopes of getting fairly on his feet again are in danger of shipwreck.

Understanding the binding force of a promise of this kind, made in writing or in the presence of witnesses, certain of the more selfish or less manly and honorable class of creditors are ever seeking to extort by fair or foul means, from an unfortunate debtor, who has honestly given up every thing, an acknowledgment of his indebtedness to them, in order that they may reap the benefit of his first efforts to get upon his feet again. Many and many an honest but indiscreet debtor has been thrown upon his back once more from this cause, and all his hopes in life blasted for ever. The means of approach to a debtor, in this situation, are many and various. "Do you think you will ever be able to do any thing on that old account?" blandly asked, in the presence of a third party, is answered by, "I hope so. But, at present, it takes every dollar I can earn for the support of my family." This is sufficient—the whole claim is in full force. In the course of a month or two, perhaps in a less period, a sheriff's writ is served, and the poor fellow's furniture, or small stock in trade, is seized, and he broken all up again. To have replied—"You have no claim against me," to the insidious question, seemed in the mind of the poor, but honest man, so much like a public confession that he was a rogue, that he could not do it. And yet this was his only right course, and he should have taken it firmly. Letters are often written, calling attention to the old matter, in which are well-timed allusions to the debtor's known integrity of character, and willingness to pay every dollar he owes in the world, if ever able. Such letters should never be answered, for the answer will be almost sure to contain something that, in a court of justice, will be construed into an acknowledgment of the entire claim. In paying off old accounts that the law has cancelled, which we think every man should do, if in his power, the acknowledgment of indebtedness never need go further than the amount paid at any time. Beyond this, no creditor, who does not wish to oppress, will ask a man to go. If any seek a further revival of the old claim, let the debtor be aware of them; and also, let him be on his guard against him who in any way alludes, either in writing or personally, to the previous indebtedness.

But we have digressed far enough. Mr. Jenkins, we are sorry to say, was not of that class of debtors who never consider an obligation morally cancelled. The law once on his side, he fully made up his mind to keep it for ever between him and all former transactions. Sundry were the attempts made to get old claims against him revived, after it was clearly understood that he was getting to be worth money; but Jenkins was a rogue at heart, and rogues are always more wary than honest men.

Among the creditors of Jenkins, was a man named Gooding, who had loaned him five hundred dollars, and lost three hundred of it—two-fifths being all that was realized from the debtor's effects. Gooding pitied sincerely the misfortunes of Jenkins, and pocketed his loss without saying a hard word, or laying the weight of a finger upon his already too heavily burdened shoulders. But it so happened, that as Jenkins commenced going up in the world, Gooding began to go down. At the time when the former was clearly worth ten thousand dollars, he was hardly able to get money enough to pay his quarterly rent-bills. Several times he thought of calling the attention of his old debtor to the balance still against him, which, as it was for borrowed money, ought certainly to be paid. But it was an unpleasant thing to remind a friend of an old obligation, and Gooding, for a time, chose to bear his troubles, as the least disagreeable of the two alternatives. At last, however, difficulties pressed so hard upon him, that he forced himself to the task.

Both he and Jenkins lived about three-quarters of a mile distant from their places of business, in a little village beyond the suburbs of the city. Gooding was lame, and used to ride to and from his store in a small wagon, which was used for sending home goods during the day. Jenkins usually walked into town in the morning, and home in the evening. It not unfrequently happened that Gooding overtook the latter, while riding home after business hours, when he always invited him to take a seat by his side, which invitation was never declined.

They were riding home in this way, one evening, when Gooding, after clearing his throat two or three times, said, with a slight faltering in his voice—

"I am sorry, neighbour Jenkins, to make any allusion to old matters, but as you are getting along very comfortably, and I am rather hard pressed, don't you think you could do something for me on account of the three hundred dollars due for borrowed money. If it had been a regular business debt, I would never have said a word about it, but"—

"Neighbour Gooding," said Jenkins, interrupting him, "don't give yourself a moment's uneasiness about that matter. It shall be paid, every dollar of it; but I am not able, just yet, to make it up for you. But you shall have it."

This was said in the blandest way imaginable, yet in a tone of earnestness.

"How soon do you think you can do something for me?" asked Gooding.

"I don't know. If not disappointed, however

I think I can spare you a little in a couple of months."

"My rent is due on the first of October. If you can let me have, say fifty dollars, then, it will be a great accommodation."

"I will see. If in my power, you shall certainly have at least that amount."

Two months rolled round, and Gooding's quarter-day came. Nothing more had been said by Jenkins on the subject of the fifty dollars, and Gooding felt very reluctant about reminding him of his promise; but he was short in making up his rent, just the promised sum. He waited until late in the day, but Jenkins neither sent nor called. As the matter was pressing, he determined to drop in upon his neighbour, and remind him of what he had said. He accordingly went round to the store of Jenkins, and found him alone with his clerk.

"How are you to-day?" said Jenkins, smiling.

"Very well How are you?"

"So, so."

Then came a pause.

"Business rather dull," remarked Jenkins.

"Very," replied Gooding, with a serious face, and more serious tone of voice. "Nothing at all doing. I never saw business so flat in my life."

"Flat enough."

Another pause.

"Ahem! Mr. Jenkins," began Gooding, after a

few moments, "do you think you can do any thing for me to-day?"

"If there is any thing I can do for you, it shall be done with pleasure," said Jenkins, in a cheerful way. "In what can I oblige you?"

"You remember, you said that in all probability you would be able to spare me as much as fifty dollars to-day?"

"*I* said so?" Jenkins asked this question with an appearance of real surprise.

"Yes. Don't you remember that we were riding home one evening, about two months ago, and I called your attention to the old account standing between us, and you promised to pay it soon, and said you thought you could spare me fifty dollars about the time my quarter's rent became due?"

"Upon my word, friend Gooding, I have no recollection of the circumstance whatever," replied Jenkins with a smile. "It must have been some one else with whom you were riding. I never said I owed you any thing, or promised to pay you fifty dollars about this time."

"Oh, yes! but I am sure you did."

"And I am just as sure that I did not," returned Jenkins, still perfectly undisturbed, while Gooding, as might be supposed, felt his indignation just ready to boil over. But the latter controlled himself as best he could; and as soon as he could get away from the store of Jenkins, without doing so in a manner that would tend to close all intercourse between them, he left and returned to his own place of business, chagrined and angry.

On the same evening, as Gooding was riding home, he saw Jenkins ahead of him on the road. He soon overtook him. Jenkins turned his usual smiling face upon his old creditor, and said, "Good evening," in his usual friendly way. The invitation to get up and ride, that was always given on like occasions, was extended again, and in a few moments the two men were riding along, side by side, as friendly, to all appearance, as if nothing had happened.

"Jenkins, how could you serve me such a scaly trick as you did?" Gooding said, soon after his neighbour had taken a seat by his side. "You know very well that you promised to pay my claim; and also promised to give me fifty dollars of it to-day, if possible."

"I know I did. But it was out of my power to let you have any thing to-day," replied Jenkins.

"But what was the use of your denying it, and making me out a liar or a fool, in the presence of your clerk?"

"I had a very good reason for doing so. My clerk would have been a witness to my acknowledgment of your whole claim against me, and thus make me liable before I was ready to pay it. As my head is fairly clear of the halter, you cannot blame me for wishing to keep it so. A burnt child, you know, dreads the fire."

"But you know me well enough to know that I never would have pressed the claim against you"

"Friend Gooding, I have seen enough of the world to satisfy me that we don't know any one. I am very ready to say to you, that your claim shall be satisfied to the full extent, whenever it is in my power to do so; but a *legal* acknowledgment of the claim I am not willing to make. You mustn't think hard of me for what I did to-day. I could not, in justice to myself, have done any thing else."

Gooding professed to be fully satisfied with this explanation, although he was not. He was very well assured that Jenkins was perfectly able to pay him the three hundred dollars, if he chose to do so, and that his refusal to let him have the fifty dollars, conditionally promised, was a dishonest act.

More than a year passed, during which time Gooding made many fruitless attempts to get something out of Jenkins, who was always on the best terms with him, but put him off with fair promises, that were never kept. These promises were never made in the presence of a third person, and might, therefore, have just as well been made to the wind, so far as their binding force was concerned. Things grew worse and worse with Gooding, and he became poorer every day, while the condition of Jenkins as steadily improved.

One rainy afternoon, Gooding drove up to the store of his old friend, about half an hour earlier than he usually left for home. Jenkins was standing in the door.

"As it is raining, I thought I would call round for you," he said, as he drew up his horse.

"Very much obliged to you, indeed," returned Jenkins, quite well pleased. "Stop a moment, until I lock up my desk, and then I will be with you."

In a minute or two Jenkins came out, and stepped lightly into the wagon.

"It is kind in you, really, to call for me," he said, as the wagon moved briskly away. "I was just thinking that I should have to get a carriage."

"It is no trouble to me at all," returned Gooding, "and if it were, the pleasure of doing a friend a kindness would fully repay it."

"You smell strong of whisky here," said Jenkins, after they had ridden a little way, turning his eyes toward the back part of the wagon as he spoke. "What have you here?"

"An empty whisky-hogshead. This rain put me in mind of doing what my wife has been teasing me to do for the last six months—get her a rain-barrel. I tried to get an old oil-cask, but couldn't find one. They make the best rain-barrels. Just burn them out with a flash of good dry shavings, and they are clear from all oily impurities, and tight as a drum."

"Indeed! I never thought of that. I must look out for one, for our old rain-hogshead is about tumbling to pieces."

From rain-barrels the conversation turned upon business, and at length Gooding brought up the old story, and urged the settlement of his claim as a matter of charity.

"You don't know how much I need it," he said. "Necessity alone compels me to press the claim upon your attention."

"It is hard, I know, and I am very sorry for you," Jenkins replied. "Next week, I will certainly pay you fifty dollars."

"I shall be very thankful. How soon after do you think you will be able to let me have the balance of the three hundred due me. Say as early as possible."

"Within three months, at least, I hope," replied Jenkins.

"Harry! Do you hear that?" said Gooding, turning his head toward the back part of the wagon, and speaking in a quick, elated manner.

"Oh, ay!" came ringing from the bunghole of the whisky-hogshead.

"Who the dickens is that?" exclaimed Jenkins, turning quickly round.

"No one," replied Gooding, with a quiet smile, "but my clerk, Harry Williams."

"Where?"

"Here," replied the individual named, pushing himself up through the loose head of the upright hogshead, and looking into the face of the discomfited Jenkins, with a broad smile of satisfaction upon his always humorous phiz.

"Whoa, Charley," said Gooding, at this moment reining up his horse before the house of Jenkins.

The latter stepped out, with his eyes upon the ground, and stood with his hand upon the wagon, in thought, for some moments; then looking up, he said, while the humour of the whole thing pressed itself so full upon him, that he could not help smiling,

"See here, Gooding, if both you and Harry will promise me never to say a word about this confounded trick, I will give you a check for three hundred dollars on the spot."

"No, I must have four hundred and twenty-six dollars, the principal and interest. Nothing less," returned Gooding firmly. "You have acknowledged the debt in the presence of Mr. Williams, and if it is not paid by to-morrow twelve o'clock, I shall commence suit against you. If I receive the money before that time, we will keep this little matter quiet; if suit is brought, all will come out on the trial."

"As you please," said Jenkins angrily, turning away, and entering his house.

Before twelve o'clock on the next day, however, Jenkins's clerk called in at the store of Gooding, and paid him four hundred and twenty-six dollars, for which he took his receipt in full for all demands to date. The two men were never afterward on terms of sufficient intimacy to ride in the same wagon together. Whether Gooding and his clerk kept the matter a secret, as they promised, we don't know. It is very certain, that it was known all over town in less than a week, and soon after was told in the newspapers, as a most capital joke.

A SHOCKING BAD MEMORY.

"MUST I give up every thing?" asked Mr. Hardy of his lawyer, with whom he was holding a consultation as to the mode and manner of getting clear of certain responsibilities in the shape of debt.

"Yes, every thing, or commit perjury. The oath you have taken is very comprehensive. If you keep back as much as ten dollars you will swear falsely."

"Bad—bad. I have about seven thousand dollars, and I owe twenty thousand. To divide this among my creditors, gives them but a small sum apiece, while it strips me of every thing. Is there no way, Mr. Dockett, by which I can retain this money, and yet not take a false oath? You gentlemen of the bar can usually find some loop-hole in the law out of which to help your clients. I know of several who have gone through the debtors' mill, and yet not come forth penniless; and some of them, I know, would not be guilty of false swearing."

"Oh yes, the thing is done every day."

"Ah, well, how is it done?"

"The process is very simple. Take your seven thousand dollars, and make it a present to some friend, in whom you can confide. Then you will be worth nothing, and go before the insolvent commissioners and swear until you are black and blue, without perjuring yourself."

"Humph! is that the way it is done?" said Mr. Hardy.

"The very way."

"But suppose the friend should decline handing it back?"

The lawyer shrugged his shoulders as he replied, "You must take care whom you trust in an affair of this kind. At worst, however, you would be just as well off, assuming that your friend should hold on to what you gave him, as you would be if you abandoned all to your creditors."

"True, if I abandon all, there is no hope of even getting back a dollar. It is the same as if I had thrown every thing into the sea."

"Precisely."

"While, in adopting the plan you propose, the chances for getting back my own again are eight to ten in my favour."

" Or, you might almost say, ten to ten. No friend into whose hands you confided the little remnant of your property would be so base as to withhold it from you."

" I will do it," said Mr. Hardy, as he parted with the lawyer.

One day, a few weeks after this interview took place, the client of Mr. Dockett came hurriedly into his office, and, drawing him aside, said, as he slipped a small package into his hand, "Here is something for you. You remember our conversation a short time ago?"

" Oh, very well."

"You understand me, Mr. Dockett?"

"Oh, perfectly! all right; when do you go before the commissioners?"

" To-morrow."

"Ah?"

" Yes—good morning. I will see you again as soon as all is over."

" Very well—good morning."

On the next day, Mr. Hardy met before the commissioners, and took a solemn oath that he had truly and honestly given up into the hands of his assignee every dollar of his property, for the benefit of his creditors, and that he did not now possess any thing beyond what the law permitted him to retain. Upon this, the insolvent commissioners gave him a full release from the claims that were held against him, and Mr. Hardy was able to say, as far as the law was concerned, "I owe no man any thing."

Mr. Dockett, the lawyer, was sitting in his office on the day after his client had shuffled off his coil of debt, his mind intent upon some legal mystery, when the latter individual came in with a light step and cheerful air.

"Good morning, Mr. Hardy," said the lawyer, smiling blandly.

"Good morning," returned the client.

"How are things progressing?" inquired the lawyer.

"All right," returned Hardy, rubbing his hands. "I am at last a free man. The cursed manacle of debt has been stricken off—I feel like a new being."

"For which I most sincerely congratulate you," returned the lawyer.

"For your kindness in so materially aiding me in the matter," said Mr. Hardy, after a pause, "I am most truly grateful. You have been my friend as well as my legal adviser."

"I have only done by you as I would have done by any other man," replied the lawyer. "You came to me for legal advice, and I gave it freely."

"Still, beyond that, you have acted as my disinterested friend," said Mr. Hardy; "and I cannot express my gratitude in terms sufficiently strong."

The lawyer bowed low, and looked just a little mistified. A slight degree of uneasiness was felt by the client. A pause now ensued. Mr. Hardy felt something like embarrassment. For some time he talked around the subject uppermost in his mind, but the lawyer did not appear to see the drift of his remarks. At last, he said—

"Now that I have every thing arranged, I will take the little package I yesterday handed you."

There was a slight expression of surprise on the countenance of Mr. Dockett, as he looked inquiringly into the face of his client.

"Handed to me?" he said, in a tone the most innocent imaginable.

"Yes," returned Hardy, with much earnestness. "Don't you recollect the package containing seven thousand dollars, that I placed in your hands to keep for me, yesterday, while I went before the commissioners?"

The lawyer looked thoughtful, but shook his head.

"Oh, but Mr. Dockett," said Hardy, now becoming excited; "you must remember it. Don't you recollect that I came in here yesterday, while you were engaged with a couple of gentlemen, and took you aside for a moment? It was then that I gave you the money."

Mr. Dockett raised his eyes to the ceiling, and mused for some time, as if trying to recall the circumstance to which allusion was made. He then shook his head, very deliberately, two or three times, remarking, as he did so, " You are evidently labouring under a serious mistake, Mr. Hardy. I have not the most remote recollection of the incident to which you refer. So far from having received the sum of money you mention, I do not remember having seen you for at least a week before to-day. I am very certain you have not been in my office within that time, unless it were when I was away. Your memory is doubtless at fault. You must have handed the money to some one else, and, in the excitement of the occasion, confounded me with that individual. Were I not charitable enough to suppose this, I should be deeply offended by what you now say."

" Mr. Dockett," returned the client, contracting his brow heavily, " Do you take me for a simpleton?"

" Pray don't get excited, Mr. Hardy," replied the lawyer, with the utmost coolness. " Excitement never does any good. Better collect your thoughts, and try and remember into whose hands you really did place your money. That I have not a dollar belonging to you, I can positively affirm."

" Perhaps you call my seven thousand dollars your own now. I gave you the sum, according to your own advice; but it was an understood matter that you were to hand the money back so soon as I had appeared before the commissioners."

" Mr. Hardy!" and the lawyer began to look angry " Mr. Hardy, I will permit neither you nor any other man to face me with such an insinuation. Do you take me for a common swindler? You came and asked if there was not some mode by which you could cheat your creditors out of six or seven thousand dollars; and I, as in duty bound, professionally, told you how the law might be evaded. And now you affirm that I joined you as a party in this nefarious transaction! This is going a little too far?"

Amazement kept the duped client dumb for some moments. When he would have spoken, his indignation was so great that he was afraid to trust himself to utter what was in his mind. Feeling that too much was at stake to enter into any angry contest with the man who had him so completely in his power, Mr. Hardy tore himself away, by a desperate effort, in order that, alone, he might be able to think more calmly, and devise, if possible, the means whereby the defective memory of the lawyer might be quickened.

On the next day, he went again to the office of his legal adviser, and was received very kindly by that individual.

"I am sure, Mr. Dockett," he said, after he was seated, speaking in a soft, insinuating tone of voice, "that you can now remember the little fact of which I spoke yesterday."

But Mr. Dockett shook his head, and answered,

"You have made some mistake, Mr. Hardy. No such sum of money was ever intrusted to me."

"Perhaps," said Hardy, after thinking for a few minutes, "I may have been in error in regard to the amount of money contained in the package. Can't you remember having received five thousand dollars from me? Think now!"

The lawyer thought for a little while, and then shook his head.

"No, I have not the slightest recollection of having received such a sum of money from you."

"The package may only have contained four thousand dollars," said Mr. Hardy, driven to this desperate expedient in the hope of inducing the lawyer to share the plunder of the creditors.

But Mr. Dockett again shook his head.

"Say, then, I gave you but three thousand dollars."

"No," was the emphatic answer.

"But I am sure you will remember having received two thousand dollars from my hand."

"No, nor one thousand, nor one hundred," replied the lawyer positively.

"Mr. Dockett, you are a knave!" exclaimed the client, springing to his feet and shaking his clenched fists at the lawyer.

"And you are both a knave and a fool," sneeringly replied Mr. Dockett.

Hardy, maddened to desperation, uttered a threat of personal violence, and advanced upon the lawyer. But the latter was prepared for him, and, before the excited client had approached three paces, there was heard a sharp click; and at the same moment, the six dark barrels of a "revolver" became visible. While Mr. Dockett thus coolly held his assailant at bay, he addressed him in this wise:

"Mr. Hardy, from what you have just said, it is clear that you have been playing a swindling game with your creditors, and stained your soul with perjury into the bargain!—Now, if you do not leave my office instantly, I will put your case in the hands of the Grand Jury, at present in session, and let you take your chance for the State prison on the charge of false swearing!"

Mr. Hardy became instantly as quiet as a lamb. For a few moments, he looked at the lawyer in bewildered astonishment, and then, turning away, left his office, in a state of mind more easily imagined than described.

Subsequently, he tried, at various times and on various occasions, to refresh the memory of Mr. Dockett on the subject of the seven thousand dollars, but the lawyer remained entirely oblivious, and to this day has not been able to recall a single incident attending the alleged transfer.

Mr. Dockett has, without doubt, a shocking bad memory.

DRIVING A HARD BARGAIN.

We know a great many business-men, famous for driving hard bargains, who would consider an insinuation that they were not influenced by honest principles in their dealings a gross outrage. And yet such an insinuation would involve only the truth. Hard bargains, by which others are made to suffer in order that we may gain, are not honest transactions; and calling them so don't in the least alter their quality.

We have our doubts whether men who overreach others in this way, are really gainers in the end. They get to be known, and are dealt with by the wary as sharpers.

A certain manufacturer—we will not say of what place, for, our story being substantially true, to particularize in this respect would be almost like pointing out the parties concerned—was obliged to use a kind of goods imported only by two or three houses. The article was indispensable in his business, and his use of it was extensive. This man, whom we will call Eldon, belonged to the class of hard bargain makers. It was a matter of principle with him never to close a transaction without, if possible, getting an advantage. The ordinary profits of trade did not satisfy him; he wanted to go a little deeper. The consequence was that almost every one was on the look out for him; and it not unfrequently happened that he paid more for an article which he imagined he was getting, in consequence of some manœuvre, at less than cost, than his next-door neighbour, who dealt fairly and above-board.

One day, a Mr. Lladd, an importer, called upon him, and said—

"I'd like to close out that entire lot of goods, Eldon. I wish you'd take them."

"How many pieces have you left?" inquired Eldon, with assumed indifference. It occurred to him, on the instant, that the merchant was a little pressed, and that, in consequence, he might drive a sharp bargain with him.

"Two hundred."

Eldon shook his head.

"What's the matter?" asked Lladd.

"The lot is too heavy."

"You'll work up every piece before six months."

"No, indeed. Not in twelve months."

"Oh, yes, you will. I looked over your account yesterday, and find that you have had a hundred and fifty pieces from me alone, and in six months."

"You must be in error."

"No. It is just as I say."

"Well, what terms do you offer?"

"If you will take the entire lot, you may have them for ten and a quarter, three months."

Eldon thought for a few moments, and then shook his head.

"You must say better than that."

"What better can you ask? You have been buying a dozen pieces at a time, for ten and a half, cash, and now I offer you the lot at ten and a quarter, three months."

"Not inducement enough. If you will say ten at six months, perhaps I will close with you."

"No. I have named the lowest price and best terms. If you like to take the goods, well and good; if not, why you can go on and pay ten and a half, cash, as before."

"I'll give you what I said."

"Oh, no, Mr. Eldon. Not a cent less will bring them."

"Very well. Then we can't trade," said the manufacturer.

"As you like," replied the merchant.

And the two men parted.

Now Eldon thought the offer of Lladd a very fair one, and meant to accept of it, if he could make no better terms; but seeing that the merchant had taken the pains to come and offer him the goods, he suspected that he was in want of money, and would take less than he asked, in order to get his note and pass it through bank. But he erred in this.

Eldon fully expected to see Mr. Lladd before three days went by. But two weeks elapsed, and as there had been no visit from the dealer, the manufacturer found it necessary to go to him, in order to get a fresh supply of goods. So he went to see him.

"I must have a dozen pieces of those goods to-day," said he, as he met Mr. Lladd.

"Very well. They are at your service."

"You'll sell them at ten and a quarter, I suppose?"

Mr. Lladd shook his head.

"But you offered them at that, you know."

"I offered the whole lot at that price, and the offer is still open; though I am in no way particular about selling."

Since ten dollars and a quarter a piece had been mentioned, the idea of paying more had become entirely obliterated from the mind of Eldon.

"But if you can sell for ten and a quarter, three months, you can sell for the same, cash."

"Yes, so I can; but I don't mean to do it." The merchant felt a little fretted.

Eldon was disappointed. He stood chaffering for some time longer; but finding it impossible to bring Lladd over to his terms, he finally agreed to take the two hundred pieces at ten and a quarter, on his note at three months.

Still he was far from being satisfied. He had fully believed that the merchant was pressed for money, and that he would in consequence be able to drive a hard bargain with him. Notwithstanding he had been compelled to go to Lladd, and to accept his terms, he yet believed that money was an object to him, and that, rather than not have the sale confirmed, he would let it be closed at ten dollars a piece, on a note at six months. So firmly was he impressed with this idea, that he finally concluded to assume, boldly, that ten dollars was the price agreed upon, and to affect surprise that the bill expressed any other rate.

In due time, the goods were delivered and the bill sent in. Immediately upon this being done, Eldon called upon the merchant and said, in a confident manner, as he laid the bill he had received upon his desk.

"You've made a mistake, haven't you?"

"How?"

"In charging these goods."

"No. I told you the price would be ten and a quarter, didn't I?"

"I believe not. I understood the terms to be ten dollars, at six months."

"You offered that, but I positively refused it."

"I am sure I understood you as accepting my offer, and ordered the goods to be sent home under that impression"

"If so, you erred," coolly replied Lladd.

"I can't take them at the price called for in this bill," said Eldon, assuming a positive air, and thinking, by doing so, Lladd would deem it his better policy to let the goods go at ten dollars.

"Then you can send them home," replied the merchant, in a manner that offended Eldon.

"Very well, I will do so, and you may keep your goods," he retorted, betraying, as he spoke, a good deal of warmth.

And the goods were sent back, both parties feeling offended; Lladd at the glaring attempt made to overreach him, and Eldon because the other would not submit to be overreached.

On the day following, Eldon started out in search of another lot of the goods he wanted, and thought himself fortunate in meeting with some in the hands of a dealer named Miller, but demurred when twelve dollars and a half a piece were asked for them.

"I can't take less," was replied.

"But," said Eldon, "Lladd has the same article for ten and a half."

"You don't pretend to put his goods alongside of mine?" returned Miller.

Eldon examined them more closely.

"They are better, it is true. But the difference is not so great as the price."

"Look again."

Another close examination was made.

"They are finer and thicker certainly. But you ask too much for them."

"It's my lowest price. They will bring it in the market, which is now bare."

"Won't you let me have a dozen pieces at twelve dollars?" asked Eldon.

"Can't sell a piece for less than what I said."

Eldon hung on for some time, but finally ordered a dozen pieces to be sent home, and paid the bill, though with a bad grace. Still, he was so angry with Lladd because he had shown a proper resentment at the effort made to overreach him, that he determined to buy no more of his goods if he could supply himself at a higher price.

Thus matters went on for five or six months, Eldon supplying himself at the store of Miller, and reconciling himself to the serious advance in price, with the reflection that Lladd's goods were remaining dead on his hands.

At last, Miller's supply was exhausted. Eldon called, one day, and ordered a dozen pieces, and received for answer—

"Not a piece in the store."

"What? All gone?" said Eldon.

"Yes, you got the last some days ago."

"I'm sorry for that. Lladd has a good stock on hand, but I don't care about dealing with him, if I can help it. He's a crusty sort of a fellow. Has no other house a supply?"

"Not to my knowledge. There is only a limited demand for the article, you know, and but few importers care about ordering it, for the reason that it goes off slowly."

Eldon tried several places, but couldn't find a yard. By the next day, his workmen would be idle; and so he had no alternative but to call upon Lladd. The merchant received him pleasantly; and they chatted for a while on matters and things in general. At last Eldon, though it went against the grain, said—

"I want you to send me twenty pieces of those goods around, with the bill."

The merchant smiled blandly and replied—

"Sorry I can't accommodate you. But I haven't a yard in the store."

"What?" Lladd looked blank.

"No. I have sold off the entire lot, and concluded not to import any more of that class of goods."

"Ah? I supposed they were still on hand."

"No, I placed them in the hands of Miller, and he has worked them all off for me at a considerable advance on former prices. He notified me, a week ago, that the lot was closed out, and rendered account sales at twelve and a half per piece."

Lladd said all this seemingly unconscious that every word he was uttering fell like a blow upon his old customer. But he understood it all very well, and had caught the hard bargain maker in a trap he little dreamed had been laid for his feet

Eldon stammered out some half coherent responses, and took his departure with more evidences of his discomfiture in his face and manner than he wished to appear. He had, in fact, been paying twelve dollars and a half for the very goods he had sent back because he couldn't get them for ten dollars, at six months credit.

Eldon did not understand how completely he had overreached himself, until a part of his establishment had been idle for days, and he had been compelled to go to New York, and purchase some fifty pieces of the goods he wanted, for cash, at twelve dollars per piece, a price that he is still compelled to pay, as neither Lladd nor any other importing house in the city has since ordered a case from abroad. So much for driving a hard bargain.

OUT OF THE FRYING-PAN INTO THE FIRE;

OR, THE LOVE OF A HOUSE.

"HADN'T you better give your landlord notice to-day, that we will move at the end of the year, Mr. Plunket?"

"Move! For heaven's sake, Sarah, what do we want to move for?"

"Mr. Plunket!"

"Mrs. Plunket!"

"It's a very strange way for you to address me, Mr. Plunket. A very strange way!"

"But for what on earth do you want to move, Sarah? Tell me that. I'm sure we are comfortable enough off here."

"Here! I wouldn't live in this miserable house another twelve months, if you gave me the rent free."

"I don't see any thing so terribly bad about the house. I am well satisfied."

"Are you, indeed! But I am not, I can tell you for your comfort."

"What's the matter with the house?"

"Every thing. There isn't a comfortable or decent room in it, from the garret to the cellar. Not one. It's a horrid place to live in; and such a neighbourhood to bring up children in!"

"You thought it a 'love of a house' a year ago."

"Me! Mr. Plunket, I never liked it; and it was all your fault that we ever took the miserable affair."

"My fault! Bless me, Sarah, what are you talking about? I didn't want to move from where we were. *I never want to move.*"

"Oh, no, you'd live in a pigstye for ever, if you once got there, rather than take the trouble to get out of it."

"Mrs. Plunket!"

"Mr. Plunket!"

Wise from experience, the gentleman deemed it better to run than fight. So, muttering to himself, he took up his hat and beat a hasty retreat.

Mrs. Plunket had a mother, a fact of which Mr. Plunket was perfectly aware, particularly as said relative was a member of his family. She happened to be present when the above spicy conversation took place. As soon as he had retired, she broke out with—"Humph! just like him; any thing to be contrary. But I wouldn't live in this old rattle-trap of a place another year for any man that ever stepped into shoe-leather. No, indeed, not I. Out of repair from top to bottom; not a single convenience, so to speak; walls cracked, paper soiled, and paint yellow as a pumpkin."

"And worse than all, ma, every closet is infested with ants and overrun with mice. Ugh! I'm afraid to open a cupboard, or look into a drawer. Why, yesterday, a mouse jumped upon me and came near going into my bosom. I almost fainted. Oh, dear! I never can live in this house another year; it is out of the question. I should die."

"No one thinks of it, except Mr. Plunket, and he's always opposed to every thing; but that's no matter. If he don't notify the landlord, we can. Live here another twelvemonth! No, indeed!"

"I saw a bill on a house in Seventh street yesterday, and I had a great mind, then, to stop and look at it. It was a beautiful place, just what we want."

"Put your things on, Sarah, right away, and go and see about it. Depend upon it, we can't dc worse than this."

"Worse! No, indeed, that's impossible. But Mr. Plunket!"

"Pshaw! never mind him; he's opposed to every thing. If you had given him his way, where would you have been now?"

Mrs. Plunket did not reply to this, for the question brought back the recollection of a beautiful little house, new, and perfect in every part, from which she had forced her husband to move, because the parlours were not quite large enough. Never, before nor since, had they been so comfortably situated.

Acting as well from her own inclination as from her mother's advice, Mrs. Plunket went and made an examination of the house upon which she had seen the bill.

"Oh, it is such a love of a house!" she said, upon her return. "Perfect in every respect: it is larger than this, and is full of closets; and the rent is just the same."

"Did you get the refusal of it?"

"Yes. I told the landlord that I would give him an answer by to-morrow morning. He says there are a great many people after it; that he could have rented it a dozen times, if he had approved the tenants who offered. He says he knows Mr. Plunket very well, and will be happy to rent him the house."

"We must take it, by all means."

"That is, if Mr. Plunket is willing."

"Willing! Of course, he'll have to be willing."

"Oh, it is such a love of a house, ma!"

"I'm sure it must be."

"A very different kind of an affair from this, you may be certain."

When Mr. Plunket came home that evening, his wife said to him, quite amiably—"Oh, you don't know what a love of a house I saw to-day up in

Seventh street; larger, better, and more convenient than this in every way, and the rent is just the same."

"But I am sure, Sarah, we are very comfortable here."

"Comfortable! Good gracious, Mr. Plunket, I should like to know what you call comfort. How can any one be comfortable in such a miserable old rattletrap of a place as this?"

"You thought it a love of a house, you remember, before we came into it."

"Me? Me? Mr. Plunket? Why, I never liked it; and it was all your fault that we ever moved here."

"My fault?"

"Yes, indeed, it was all your fault. I wanted the house in Walnut street, but you were afraid of a little more rent. Oh, no, Mr. Plunket, you mustn't blame me for moving into this barracks of a place; you have only yourself to thank for that; and now I want to get out of it on the first good opportunity."

Poor Mr. Plunket was silenced. The very boldness of the position taken by his wife completely knocked him *hors du combat.* His fault, indeed! He would have lived on, year after year, in a log cabin, rather than encounter the horrors of moving; and yet he was in the habit of moving about once a year. What could he do now? He had yielded so long to his wife, who had grown bolder at each concession, that opposition was now hopeless. Had she stood alone, there might have been some chance for him; but backed up, as she was, by her puissant mother, victory was sure to perch on her banner; and well did Mr. Plunket know this.

"It will cost at least a hundred and fifty or two hundred dollars to move," he ventured to suggest.

"Indeed, and it will cost no such thing. I'll guaranty the whole removal for ten dollars"

"It cost over a hundred last year."

"Nonsense! it didn't cost a fifth of it."

But Mr. Plunket knew—he had the best right to know, for he had paid the bills.

From the first, Mr. Plunket felt that opposition was useless. A natural repugnance to change and a horror of the disorder and discomfort of moving caused him to make a feeble resistance; but the opposing current swept strongly against him, and he had to yield.

The house in Seventh street was taken, and, in due time, the breaking up and change came. Carpets were lifted, boxes, barrels, and trunks packed, and all the disorderly elements of a regular moving operation called into activity. Every preparation had been made on the day previous to the contemplated flight; the cars were to be at the door by eight o'clock on the next morning. In anticipation of this early movement, the children had been dragged out of bed an hour before their usual time for rising. They were, in consequence, cross and unreasonable; but not more so than mother, grandmother, and nurse, all of whom either boxed them, scolded them, or jerked them about in a most violent manner. Breakfast was served early; but such a breakfast! the least said about that the better. It was well there were no keen appetites to turn away with disappointment.

"Strange that the cars are not here!" said Mr. Plunket, who had put himself in going order. "It's nearly half an hour past the time now. Oh, dear! confound all this moving, say I."

"That's a strange way for you to talk before children, Mr. Plunket," retorted his wife.

"And this is a much stranger way for you to act, madam; for ever dragging your husband and children about from post to pillar. For my part, I feel like Noah's dove, without a place to rest the sole of my foot."

"Mr. Plunket!"

"Mrs. Plunket!"

A war of words was about commencing, but the furniture-cars drove up at the moment, when an armistice took place.

In due time, the family of the Plunkets were, bag and baggage, in their new house. A lover of quiet the male head of the establishment tried to refrain from any remarks calculated to excite his helpmate, but this was next to impossible, there being so much in the new house that he could not, in conscience, approve. If Mrs. Plunket would have kept quiet, all might have gone on very smoothly; but Mrs. Plunket could not or would not keep quiet. She was extravagant in her praise of every thing, and incessant in her comparisons between the old and the new house. Mr. Plunket listened, and bit his lip to keep silent. At last the lady said to him, with a coaxing smile, for she was not going to rest until some words of approval were extorted from her liege lord—"Now, Mr. Plunket, don't you think this a love of a house?"

"No!" was the gruff answer.

"Mr. Plunket! Why, what is your objection? I'm sure we can't be more uncomfortable than we have been for a year."

"Oh, yes, we can."

"How so?"

"There is such a thing as going from the frying-pan into the fire."

"Mr. Plunket!"

"Just what you'll find we have done, madam."

"How will you make that appear, pray?"

"In a few words. Just step this way. Do you see that building?"

"I do."

"Just to the south-west of us; from that quarter the cool breezes of summer come. We shall now have them fragrant with the delightful exhalations of a slaughter-house. Humph! Won't that be delightful? Then, again, the house is damp."

"Oh, no. The landlord assured me it was as dry as a bone."

"The landlord lied, then. I've been from garret to cellar half a dozen times, and it is just as I say. My eyes never deceive me. As to its being a better or more comfortable house, that is all in my eye. I wouldn't give as much for it, by fifty dollars, as for the one we have left."

Notwithstanding Mrs. Plunket's efforts to induce her husband to praise the house, she was not as well satisfied with it as she was at the first inspection of the premises.

"I'm sure," she replied, in rather a subdued manner, "that it is quite as good as the old house, and has many advantages over it."

"Name one," said her husband.

"It is not overrun with vermin."

"Wait a while and see."

"Oh, I know it isn't."

"How do you know?"

"I asked the landlord particularly."

"And he said no?"

"He did."

"Humph! We shall see."

And they did see. Tired out with a day's moving and fixing, the whole family, feeling hungry, out of humour, and uncomfortable, descended to the kitchen, after it had become dark, to overhaul the provision-baskets, and get a cold cut of some kind. But, alas! to their dismay, it was found that another family, and that a numerous one, already had possession. Floor, dresser, and walls were alive with a starving colony of enormous cockroaches, and the baskets, into which bread, meats, &c. had been packed, were literally swarming with them.

In horror, man, woman, and child beat a hasty retreat, and left the premises.

It would hardly be fair to record all the sayings and doings of that eventful evening. Overwearied in body and mind, the family retired to rest, but some of them, alas! not to sleep. From washboards and every other part of the chamber in which a crevice existed, crept out certain little animals not always to be mentioned to ears polite, and, more bold than the denizens of the kitchen, made immediate demonstrations on the persons of master, mistress, child, and maid.

It took less than a week to prove satisfactorily to Mrs. Plunket, though she did not admit the fact, that the new house was not to be compared with the old one in any respect. It had not a single advantage over the other, while the disadvantages were felt by every member of the family.

In a few months, however, Mr. Plunket began to feel at home, and to settle down into contentment but as he grew better and better satisfied, his wife grew more and more desirous of change, and is now, as the year begins to draw to a close, looking about her for bills on houses, and examining, every day, the "to let" department of the newspapers with a lively degree of interest. Mr. Plunket will, probably, resist stoutly when his lady proposes some new "love of a house," but it will be of no use; he will have to pull up stakes and try it again. It is his destiny; he has got a moving wife, and there is no help for him.

MARRYING A COUNT.

"Is any body dead?"

"Yes, somebody dies every second."

"So they say. But I don't mean that. Why are you looking so solemn?"

"I am not aware that I look so very solemn."

"You do, then, as solemn as the grave."

"Then I must be a grave subject." The young man affected to smile.

"You smile like a death's head, Abel. What is the matter?"

Abel Lee took his interrogator by the arm, and drew him aside. When they were a little apart from the company, he said in a low voice—

"You know that I have taken a fancy to Arabella Jones?"

"Yes, you told me that a month ago."

"She is here to-night."

"So I see."

"And is as cold to me as an icicle."

"For a very plain reason."

"Yes, too plain."

"Whiskers and moustaches are driving all before them. The man is nothing now; hair is every thing. Glover will carry off the prize unless you can hit upon some plan to win back the favour of Miss Arabella. You must come forward with higher attractions than this rival can bring."

Lee drew his fingers involuntarily over his smooth lip and chin, a movement which his friend observed and comprehended.

"Before the hair can grow Arabella will be won," he said.

"Do you think I would make such a fool of myself."

"Fool of yourself! What do you mean by that? You say you love Arabella Jones. If you wish to win her, you must make yourself attractive in her eyes. To make yourself attractive, you have only to cultivate whiskers, moustaches, and an imperial, and present a more luxuriant crop than Glover. The whole matter is very simple, and comprised in a nut-shell. The only difficulty in the way is the loss of time consequent upon the raising of this hairy crop. It is plain, in fact, that you must take a shorter way; you must purchase what you haven't time to grow. Hide yourself for a week or two, and then make your appearance with enough hair upon your face to conceal one-half or two-thirds of your features, and your way to the heart of Miss Jones is direct."

"I feel too serious on the subject to make it a matter of jesting," said Lee, not by any means relishing the levity of his friend.

"But, my dear sir," urged the friend, "what I propose is your only chance. Glover will have it all his own way, if you do not take some means to head him off. The matter is plain enough. In the days of chivalry, a knight would do almost any unreasonable thing—enter upon almost any mad adventure—to secure the favour of his lady-love; and will you hesitate when nothing of more importance than the donning of false whiskers and moustaches is concerned? You don't deserve to be thought of by Miss Jones."

"Jest away, Marston, if it is so pleasant to you," remarked Lee, with a slightly offended air.

"No, but my dear fellow, I am in earnest. I really wish to serve you. Still if the only plan at all likely to succeed is so repugnant to your feelings, you must let the whole matter go. Depend upon it, there is no other chance for you with the lady."

"Then she must go. I would not make a fool of myself for the Queen of Sheba. A man who sacrifices his own self-respect in order to secure the love of a woman becomes unworthy of her love."

"Well said, Abel Lee! That is the sentiment of a right mind, and proves to me that Arabella Jones is unworthy of you. Let her go to the whiskers, and do you try to find some one who has soul enough to love the man."

The young men separated, to mingle with the company. Marston could not help noticing Miss Arabella Jones more particularly than before, and perceived that she was coldly polite to all the young men who ventured to approach her, but warm and smiling as a June morning to an individual named Glover who had been abroad and returned home rich in hairy honours, if in nothing else. The manners of this Glover distinguished him as much as his appearance.

"To think that a woman could be attracted by a thing like that!" he said to himself a little pettishly, as he saw the alacrity with which Arabella seized the offered arm of Glover to accompany him to the supper table.

Marston was a fellow of a good deal of humour,

and relished practical joking rather more than was consistent with the comfort of other people. We cannot commend him for this trait of character. But it was one of his faults, and all men have their failings. It would have given him great pleasure, could he have induced Abel Lee to set up a rivalry in the moustache and whisker line; but Abel had too much good sense for that, and Marston, be it said to his credit, was rejoiced to find that he had. Still, the idea having once entered his head, he could not drive it away. He had a most unconquerable desire to see some one start in opposition to Glover, and was half tempted to do it himself, for the mere fun of the thing. But this was rather more trouble than he wished to take.

Not very long after this, a young stranger made his appearance in fashionable circles, and created quite a flutter among the ladies. He had, besides larger whiskers, larger moustache, and larger imperial than Glover, a superb goatee, and a decided foreign accent. He soon threw the American in the shade, especially as a whisper got out that he was a French count travelling through the country, who purposely concealed his title. The object of his visit, it was also said, was the selection of a wife from among the lovely and unsophisticated daughters of America. He wished to find some one who had never breathed the artificial air of the higher circles in his own country; who would love him for himself alone, and become his loving companion through life.

How all these important facts in relation to him got wind few paused to inquire. Young ladies forgot their plain-faced, untitled, vulgar lovers, and put on their best looks and most winning graces for the count. For a time he carried all before him. Daily might he be seen in Chestnut street, gallanting some favoured belle, with the elegant air of a dancing-master, and the grimace of a monkey. Staid citizens stopped to look at him, and plain old ladies were half in doubt whether he were a man or a pongo.

At last the count's more particular attentions were directed toward Miss Arabella Jones, and from that time the favoured Glover found that his star had passed its zenith. It was in vain that he curled his moustache more fiercely, and hid his chin in a goatee fully as large as the count's; all was of no avail. The ladies generally, and Miss Arabella in particular, looked coldly upon him.

As for Abel Lee, the bitterness of his disappointment was already past. The conduct of Arabella had disgusted him, and he therefore looked calmly on and marked the progress of events.

At length the count, from paying marked attention to Arabella in company, began to visit her occasionally at her father's house, little to the satisfaction of Mr. Jones, the father, who had never worn a whisker in his life, and had a most bitter aversion to moustaches. This being the case, the course of Arabella's love did not, it may be supposed, run very smooth, for her father told her very decidedly that he was not going to have "that monkey-faced fellow" coming about his house. Shocked at such vulgar language, Arabella replied—

"Gracious me, father! Don't speak in that way of Mr. De Courci. He's a French count, travelling in disguise."

"A French monkey! What on earth put that nonsense into your head?"

"Everybody knows it, father. Mr. De Courci tried to conceal his rank, but his English valet betrayed the secret. He is said to be connected with one of the oldest families in France, and to have immense estates near Paris."

"The largest estates he possesses are in Whiskerando, if you ever heard of that place. A French count! Preposterous!"

"I know it to be true," said Arabella, emphatically.

"How do you know it, Miss Confidence?"

"I know it from the fact that I hinted to him, delicately, my knowledge of his rank abroad, and he did not deny it. His looks and his manner betrayed what he was attempting to conceal."

"Arabella!" said Mr. Jones, with a good deal of sternness, "if you were silly enough to hint tc this fellow what you say you did, and he was im postor enough not to deny it on the spot in the most unequivocal terms, then he adds the character of a designing villain to that of a senseless fop. In the name of homely American common sense, can you not see, as plain as daylight, that he is no nearer akin to a foreign nobleman than his barber or boot-black may be?"

Arabella was silenced because it was folly to contend in this matter with her father, who was a blunt, common-sense, clear-seeing man; but she was not in the least convinced Mr. De Courci was not a French count for all he might say, and, what was better, evidently saw attractions in her superior to those of which any of her fair compeers could boast.

"My dear Miss Jones," said the count, when they next met, speaking in that delightful foreign accent, so pleasant to the ear of the young lady, and with the frankness peculiar to his nature; "I cannot withhold from you the honest expression of my sentiments. It would be unjust to myself, and unjust to you; for those sentiments too nearly involve my own peace, and, it may be, yours."

The count hesitated, and looked interesting. Arabella blushed and trembled. The words, "You will speak to my father," were on the young lady's tongue. But she checked herself, and remained silent. It would not do to make that reference of the subject.

Then came a gentle pressure of hair upon her cheek, and a gentle pressure from the gloved hand in which her own was resting.

"My dear young lady, am I understood?"

Arabella answered, delicately, by returning the gentle pressure of her hand, and leaning perceptibly nearer the Count De Courci.

"I am the happiest of men!" said the count, enthusiastically.

"And I the happiest of women," responded Arabella, not audibly, but in spirit.

"Your father?" said De Courci. "Shall I see him?"

"It will not be well yet," replied the maiden, evincing a good deal of confusion. "My father is"—

"Is what?" asked the nobleman, slightly elevating his person.

"Is a man of some peculiar notions. Is, in fact, too rigidly American. He does not like"—

Arabella hesitated.

"Doesn't like foreigners. Ah! I comprehend," and the count shrugged his shoulders and looked dignified; that is, as dignified as a man whose face is covered with hair can look.

"I am sorry to say that he has unfounded prejudices against every thing not vulgarly American."

"He will not consent, then?"

"I fear not, Mr. De Courci."

"Hum-m. Ah!" and the count thought for some

moments. "Will not consent. What then? Arabella!" and he warmed in his manner—"Arabella, shall an unfounded prejudice interpose with its icy barriers? Shall hearts that are ready to melt into one, be kept apart by the mere word of a man? Forbid it, love! But suppose I go to him?"

"It will be useless! He is as unbending as iron."

Such being the case, the count proposed an elopement, to which Arabella agreed, after the expression of as much reluctance as seemed to be called for.

A few weeks subsequently, Mr. Jones received a letter from some person unknown, advising him of the fact that if at a certain hour on that evening he would go to a certain place, he would intercept Mr. De Courci in the act of running away with his daughter. This intelligence half maddened the father. He hurried home, intending to confront Arabella with the letter he had received, and then lock her up in her room. But she had gone out an hour before. Pacing the floor in a state of strong excitement, he awaited her return until the shadows of evening began to fall. Darkness closed over all things, but still she was away, and it soon became evident that she did not mean to come back.

It was arranged between De Courci and Arabella that he was to wait for her with a carriage at a retired place in the suburbs, where she was to join him. They were then to drive to a minister's, get the marriage ceremony performed, and proceed thence to take possession of an elegant suite of rooms which had been engaged in one of the most fashionable hotels in the city. To escape all danger of interference with her movements, the young lady had left home some hours before evening, and spent the time between that and the blissful period looked for with such trembling delight, in the company of a young friend and confidante. Darkness at length threw a veil over all things, and under cover of this veil Arabella went forth alone, and hurried to the appointed place of meeting. A lamp showed her the carriage in waiting, and a man pacing slowly the pavement near by, while she was a considerable distance off. Her heart beat wildly, the breath came heavily up from her bosom. She quickened her pace, but soon stopped suddenly in alarm, for she saw a man advancing rapidly from another quarter. It a few moments this individual came up to the person who was walking before the carriage, and whom she saw to be her lover. Loud words instantly followed, and she was near enough to hear an angry voice say—

"I'll count you, you base scoundrel!"

It was the voice of her father! Fearful lest violence should be done to her lover, Arabella screamed and flew to the spot. Already was the hand of Mr. Jones at De Courci's throat, but the count in disguise, not relishing the rough grasp of the indignant father, disengaged himself and fled ingloriously, leaving poor Arabella to the unbroken fury cf his ire. Without much ceremony he thrust her into the waiting carriage, and, giving the driver a few hurried directions, entered himself. What passed between the disappointed countess, that was to be, and her excited father, it is not our business to relate.

Not content with having interrupted this nice little matrimonial arrangement, Mr. Jones called at the hotel where De Courci put up, early on the next morning. But the elegant foreigner had not occupied his apartments during the night. He called a few hours later, but he had not yet made his appearance; in the morning, but De Courci was still away. On the next morning the following notice appeared in one of the daily newspapers:—

"NIPPED IN THE BUD.—Fashionable people will remember a whiskered, mustachioed fellow with a foreign accent, named De Courci, who has been turning the heads of half the silly young girls in town for the last two months. He permitted it to leak out, we believe, that he was a French count, with immense estates near Paris, who had come to this country in order to look for a wife. This was of course believed, for there are people willing to credit the most improbable stories in the world. Very soon a love affair came on, and he was about running off with the silly daughter of a good substantial citizen. By some means the father got wind of the matter, and repaired to the appointed place of meeting just in time. He found De Courci and a carriage in waiting. Without much ceremony, he laid violent hands on the count, who thought it better to run than to fight, and therefore fled ingloriously, just as the daughter arrived on the ground. He has not been heard of since. We could write a column by way of commentary upon this circumstance, but think that the facts in the case speak so plainly for themselves, that not a single remark is needed to give them force. We wish the lady joy at her escape, for the count in disguise is no doubt a scheming villain at heart."

Poor Arabella was dreadfully cut down when this notice met her eye. It was a long time before she ventured into company again, and ever after had a mortal aversion to mustaches and imperials. The count never after made his appearance in Philadelphia.

The young man named Marston, who had jested with Abel Lee about the loss of his lady-love, was seated in his room some ten minutes after the sudden appearance of Mr. Jones at the place of meeting between the lovers, when his door was thrown open, and in bounded De Courci, hair and all! Cloak, hat, and hair were instantly thrown aside, and a smooth, young, laughing face revealed itself from behind whiskers, moustaches, imperials, and goatee

"Where's the countess?" asked Marston, in a merry voice. "Did she faint?"

"Dear knows. That sturdy old American father of hers got me by the throat before I could say Jack Robinson, and I was glad to make off with a whole skin. Arabella arrived at the moment, and gave a glorious scream. Of any thing further, deponent sayeth not."

"She'll be cured of moustaches, or I'm no prophet."

"I guess she will. But the fact is, Marston," and the young man looked serious, "I'm afraid this joke has been carried too far."

"Not at all. The moral effect will tell upon our silly young ladies, whose heads are turned with a foreign accent and a hairy lip. You acted the whiskered fop to a charm. No one could have dreamed that all was counterfeit."

"So far as the general effect is concerned, I have no doubt; but I'm afraid it was wrong to victimize Miss Arabella for the benefit of the whole race of weak-minded girls. The effect upon her may be more serious than we apprehend."

"No, I think not. The woman who could pass by as true a young man as Abel Lee for a foreign count in disguise, hasn't heart enough to receive a deep injury. She will be terribly mortified, but that will do her good."

"If it turn out no worse than that, I shall be glad. But I must own, now that the whole thing is over, that I am not as well satisfied with myself as I thought I would be. I don't know what my good sisters at the South would say, if they knew I had been engaged in such a mad-cap affair. But I lay all the blame upon you. You, with your cool head, ought to have known better than to start a young hot-brained fellow like me, just let loose from college, upon such a wild adventure. I'm afraid that if Jones had once got me fairly into his clutches, he would have made daylight shine through me."

"Ha! ha! No doubt of it. But come, don't begin to look long-faced. We will keep our own counsel, and no one need be the wiser for our participation in this matter. Wait a while, and let us enjoy the nine days' wonder that will follow."

But the young man, who was a relative of Marston, and who had come to the city fresh from college, just in the nick of time for the latter, felt, now that the excitement of his wild prank was over, a great deal more sober about the matter than he had expected to feel. Reason and reflection told him that he had no right to trifle with any one as he had trifled with Arabella Jones. But it was too late to mend the matter. No great harm, however, came of it; and perhaps, good; for a year subsequently, Abel Lee conducted his old flame to the altar, and she makes him a loving and faithful wife.

JOB'S COMFORTERS;

OR, THE LADY WITH NERVES.

WHAT a blessed era in the world's history that was when the ladies had no nerves! Alas! I was born too late instead of too early, as the complaint of some is. I am cursed with nerves, and, as a consequence, am ever and anon distressed with nervous fears of some direful calamity or painful affliction. I am a simpleton for this, I know; but then, how can I help it? I try to be a woman of sense, but my nerves are too delicately strung. Reason is not sufficient to subdue the fears of impending evil that too often haunt me.

It would not be so bad with me, if I did not find so many good souls ready to add fuel to the flames of my fears. One of my most horrible apprehensions, since I have been old enough to think about it, has been of that dreadful disease, cancer. I am sure I shall die of it,—or, if not, some time in life have to endure a frightful operation for its removal.

I have had a dull, and sometimes an acute pain

in one of my breasts, for some years. I am sure it is a cancer forming, though my husband always ridicules my fears. A few days ago a lady called in to see me. The pain had been troubling me, and I felt nervous and depressed.

"You don't look well," said my visitor.

"I am not very well," I replied.

"Nothing serious, I hope?"

"I am afraid there is, Mrs. A——." I looked gloomy, I suppose, for I felt so.

"You really alarm me. What can be the matter?"

"I don't know that I have ever mentioned it to you, but I have, for a long time, had a pain in my left breast, where I once had a gathering, and in which hard lumps have ever since remained. These have increased in size, of late, and I am now confirmed in my fears that a cancer is forming."

"Bless me!" And my visitor lifted both hands and eyes. "What kind of a pain is it?"

"A dull, aching pain, with occasional stitches running out from one spot, as if roots were forming."

"Just the very kind of pain that Mrs. N—— had for some months before the doctors pronounced her affection cancer. You know Mrs. N——?"

"Not personally. I have heard of her."

"You know she had one of her breasts taken off?"

"Had she?" I asked, in a husky voice. I had horrible feelings.

"Oh, yes!" My visitor spoke with animation.

"She had an operation performed about six months ago. It was dreadful! Poor soul!"

My blood fairly curdled; but my visitor did not notice the effect of her words.

"How long did the operation last?" I ventured to inquire.

"Half an hour."

"Half an hour! So long?"

"Yes; it was a full half hour from the time the first incision was made until the last little artery was taken up."

"Horrible! horrible!" I ejaculated, closing my eyes, and shuddering.

"If so horrible to think of, what must it be in reality?" said my thoughtless visitor. "If it were my case, I would prefer death. But Mrs. N—— is not an ordinary woman. She possesses unusual fortitude, and would brave any thing for the sake of her husband and children. It took even her, however, a long time to make up her mind to have the operation performed; and it was only when she was satisfied that further delay would endanger her life, that she consented to have it done. I saw her just the day before; she looked exceedingly pale, and said but little. A very intimate friend was with her, whom I was surprised to hear talk to her in the liveliest manner, upon subjects of the most ordinary interest. She was relating a very amusing story which she had read, when I entered, and was laughing at the incidents. Even Mrs. N—— smiled. It seemed to me very much out of place, and really a mockery to the poor creature; it was downright cruel How any one could do so I cannot imagine. 'My dear madam,' I said as soon as I could get a chance to speak to her, 'how do you feel? I am grieved to death at the dreadful operation you will have to go through. But you must bear it bravely; it will soon be over.' She thanked me with tears in her eyes for my kind sympathies, and said that she hoped she would be sustained through the severe trial. Before I could get a chance to reply, her friend broke in with some nonsensical stuff that made poor Mrs. N—— laugh in spite of herself, even though the tears were glistening on her eyelashes. I felt really shocked. And then she ran on in the wildest strain you ever heard, turning even the most serious remark I could make into fun. And, would you believe it? she treated with levity the operation itself whenever I alluded to it, and said that it was nothing to fear—a little smarting and a little pain, but not so bad as a bad toothache, she would wager a dollar.

"'That is all very well for you to say,' I replied, my feelings of indignation almost boiling over, 'but if you had the operation to bear, you would find it a good deal worse than a bad toothache, or the severest pain you ever suffered in your life.'

"Even this was turned into sport. I never saw such a woman. I believe she would have laughed in a cholera hospital. I left, assuring Mrs. N—— of my deepest sympathies, and urged her to nerve herself for the sad trial to which she was so soon to be subjected. I was not present when the operation was performed, but one who attended all through the fearful scene gave me a minute description of every thing that occurred."

The thought of hearing the details of a dreadful operation made me sick at heart, and yet I felt a morbid desire to know all about it. I could not ask my visitor to pause; and yet I dreaded to hear her utter another sentence. Such was the strange disorder of my feelings! But it mattered not what process of thought was going on in my mind, or what was the state of my feelings; my visitor went steadily on with her story, while every fifth word added a beat to my pulse per minute.

The effect of this detail was to increase all the cancerous symptoms in my breast, or to cause me to imagine that they were increased. When my husband came home, I was in a sad state of nervous excitement. He anxiously inquired the cause.

"My breast feels much worse than it has felt for a long time," said I. "I am sure a cancer is forming. I have all the symptoms."

"Do you know the symptoms?" he asked.

"Mrs. N—— had a cancer in her breast, and my symptoms all resemble hers."

"How do you know?"

"Mrs. A—— has been here, and she is quite intimate with Mrs. N——. All my symptoms, she says, are precisely like hers."

"I wish Mrs. A—— was in the deserts of Arabia!" said my husband, in a passion. "Even if what she said were true, what business had she to say it? Harm, not good, could come of it. But I don't believe you have any more cancer in your breast than I have. There is an obstruction and hardening of the glands, and that is about all."

"But Mrs. N——'s breast was just like mine, for Mrs. A—— says so. She described the feeling Mrs. N—— had, and mine is precisely like it."

"Mrs. A—— neither felt the peculiar sensation in Mrs. N——'s breast nor in yours; and, therefore, cannot know that they are alike. She is an idle, croaking gossip, and I wish she would never cross our threshold. She always does harm."

I felt that she had done me harm, but I wouldn't say so. I was a good deal vexed at the way my husband treated the matter, and accused him of indifference as to whether I had a cancer or not. He bore the accusation very patiently, as, indeed, he always does any of my sudden ebullitions of feeling. He knows my weakness.

"If I thought there were danger," he mildly said, "I would be as much troubled as you are."

"As to danger, that is imminent enough," I returned, fretfully.

"On the contrary, I am satisfied that there is none. One of your symptoms makes this perfectly clear."

"Indeed! What symptom?" I eagerly asked.

"Your terrible fears of a cancer are an almost certain sign that you will never have one. The evil we most fear, rarely, if ever, falls upon us."

"That is a very strange way to talk," I replied.

"But a true way, nevertheless," said my husband.

"I can see no reason in it. Why should we be troubled to death about a thing that is never going to happen?"

"The trouble is bad enough, without the reality, I suppose. We are all doomed to have a certain amount of anxiety and trouble here, whether real or imaginary. Some have the reality, and others the imagination. Either is bad enough; I don't know which is worse."

"I shall certainly be content to have the imaginary part," I replied.

"That part you certainly have, and your full share of it. I believe you have, at some period or other, suffered every ill that flesh is heir to. As for me, I would rather have a good hearty fit of sickness, a broken leg or arm, or even a cancer, and be done with it, than become a living Pandora's box, even in imagination."

"As you think I am?"

"As I know you are"

"Then you would really like to see me have a cancer in my breast, and be done with it?" I said this pretty sharply.

"Don't look so fiercely at me," returned my husband, smiling. "I didn't say I would rather *you* would have a cancer; I said *I* would rather have one, and be done with it, than suffer as you do from the fear of it, and a hundred other evils."

"I must say you are quite complimentary to your wife," I returned, in a little better humour than I had yet spoken. The fact was, my mind took hold of what my husband said about real and imaginary evils, and was somewhat braced up. Of imaginary evils I had certainly had enough to entitle me to a whole lifetime exemption from real ones.

From the time Mrs. A—— left me until my husband came in, the pain in my breast had steadily increased, accompanied by a burning and stinging sensation. In imagination, I could clearly feel the entire cancerous nucleus, and perceive the roots eating their way in all directions around it. This feeling, when I now directed my thoughts to my breast, was gone—very little pain remained.

After tea, my husband went out and returned in about an hour. He said he had been round to consult with our physician, who assured him that he had seen hundreds of cases like mine, not one of which terminated in cancer; that such glandular obstructions were common, and might, under certain circumstances, unless great care were used, cause inflammation and suppuration; but were no more productive of cancer, a very rare disease, and consequent upon hereditary tendencies, than were any of the glandular obstructions or gatherings in other parts of the body.

"But the breast is so tender a place," I said.

"And yet," returned my husband, "the annals of surgery show ten cancers in other parts of the body to one in the breast."

In this way my husband dissipated my fears, and restored my mind to a comparatively healthy state. This, however, did not long remain; I was attacked on the next day with a dull, deeply-seated pain in one of my jaw-teeth. At first, I did not regard it much, but its longer continuance than usual began to excite my fears, especially as the tooth was, to all appearance, sound.

While suffering from this attack, I had a visit from another friend of the same class with Mrs. A——. She was a kind, good-natured soul, and would watch by your sick-bed untiringly, night after night, and do it with real pleasure. But she had, like Mrs. A——, a very thoughtless habit of relating the many direful afflictions and scenes of human suffering it had been her lot to witness and hear of, unconscious that she often did great harm thereby, particularly when these things were done, as was too often the case, *apropos*.

"You are not well," she said, when she came in and saw the expression of pain in my face. "What is the matter?"

"Nothing more than a very troublesome toothache," I replied.

"Use a little kreosote," said she.

"I would; but the tooth is sound."

"A sound tooth, is it?" My visitor's tone and look made my heart beat quicker.

"Yes, it is perfectly sound."

"I am always afraid of an aching tooth that is perfectly sound, since poor Mrs. P—— had such a time with her jaw."

"What was that?" I asked, feeling instantly alarmed.

"Which tooth is it that aches?" my friend asked.

I pointed it out.

"The very same one that troubled Mrs. P—— for several months, night and day."

"Was the pain low and throbbing?" I eagerly asked.

"Yes; that was exactly the kind of pain she had."

"And did it continue so long as several months?"

"Oh, yes. But that wasn't the worst! the aching was caused by the formation of an abscess."

"A what?" A cold chill passed over me.

"An abscess."

"At the root of her tooth?"

"Yes. But that wasn't so bad as its consequences; the abscess caused the bone to decay, and produced what the doctors called a disease of the antrum, which extended until the bone was eaten clear through, so that the abscess discharged itself by the nostrils."

"Oh, horrible!" I exclaimed, feeling as sick as death, while the pain in my tooth was increased fourfold. "How long did you say this abscess was in forming?"

"Some months."

"Did she have an operation performed?" I have a terrible fear of operations.

"Oh, yes. It was the only thing that saved her life. They scraped all the flesh away on one cheek and then cut a hole through the bone. This was after the tooth had been drawn, in doing which the jaw-bone was broken dreadfully. It was months before it healed, or before she could eat with any thing but a spoon."

This completely unmanned, or, rather, unwomanned me. I asked no more questions, although my visitor continued to give me a good deal of minute information on the subject of abscesses, and the dreadful consequences that too frequently attended them. After she left another friend called, to whom I mentioned the fact of having a very bad tooth-ache, and asked her if she had ever known any one to have an abscess at the root of a sound tooth

She replied that tooth-ache from that cause was not unfrequent, and that, sometimes, very bad consequences resulted from it She advised me, by all means, to have the tooth extracted.

"I can't bear the thought of that," I replied. "I never had but one tooth drawn, and when I think of having another extracted I grow cold all over."

"Still that is much better than having caries of the jaw, which has been known to attend an abscess at the root of a tooth."

"But this does not always follow?"

"No. It is of rare occurrence, I believe. Though no one knows when such a disease exists, nor where it is going to terminate. Even apart from caries of the jaw, the thing is painful enough. Mrs. T——, an intimate friend of mine, suffered for nearly a month, night and day, and finally had to have the tooth extracted, when her mouth was so much inflamed, and so tender, that the slightest touch caused the most exquisite pain. A tumor was found at the root of the tooth as large as a pigeon's egg!"

This completed the entire overthrow of my nerves. I begged my friend, in mercy to spare me any further relations of this kind. She seemed half offended, and I had to explain the state of mind which had been produced by what a former visitor had said She, evidently, thought me a very weak woman No doubt I am.

"In the dumps again, Kate?" said my husband, when he returned home in the evening. "What is the matter now?"

"Enough to put you or any one else in the dumps," I replied fretfully. "This tooth-ache grows worse, instead of better."

"Does it, indeed? I am really very sorry. Can't any thing be done to relieve you?"

"Nothing, I am persuaded. The tooth is sound, and there must be an abscess forming at the root, to occasion so much pain."

"Who, in the name of common sense, has put this in your head?"

My husband was worried.

"Has Mrs. A—— been here again?"

"No," was my simple response.

"Then what has conjured up this bugbear to frighten you out of your seven senses?"

I didn't like this language at all. My husband seemed captious and unreasonable. Dear soul! I supposed he had cause; for they say a nervous woman is enough to worry a man's life out of him; and, dear knows, I am nervous enough! But I had only my fears before me then: I saw that my husband did not sympathize with me in the least. I merely replied—

"It may be very well for you to speak to your wife in this way, after she has suffered for nearly three days with a wretched tooth-ache If the tooth were at all decayed, or there were any apparent cause for the pain, I could bear it well enough, and wouldn't trouble you about it. But it is so clear to my mind now, that nothing but a tumour forming at the root could produce such a steady, deep-seated, throbbing pain, that I am with reason alarmed; and, instead of sympathy from my husband I am met with something very much like ridicule."

"My dear Kate," said my husband, tenderly, and in a serious voice, "pardon my apparent harshness and indifference. If you are really so serious about the matter, it may be as well to consult a dentist, and get his advice. He may be able to relieve very greatly your fears, if not the pain in your jaw."

"He will order the tooth to be extracted, I have not the least doubt."

"If there should be a tumour at the root, it will be much safer to have it out than let it remain."

A visit to the dentist *at once* was so strenuously urged by my husband, that I couldn't refuse to go. I got myself ready, and we went around before tea. I did not leave the house, however, before making my husband promise he would not insist upon my having the tooth taken out on the first visit. This he did readily.

The dentist, after examining very carefully the tooth pointed out to him, said that he didn't believe *that* tooth ached at all.

"Not ache, doctor?" said I, a little indignantly

"If you had it in your head, you would think it ached."

"Pardon me, madam," he returned, with a polite bow. "I did not mean to say that you were not in pain. I only mean to say that I think that you are mistaken in its exact locality."

"I don't see how I can be. I have had it long enough, I should think, to determine its locality with some certainty."

"Let me examine your mouth again, madam," said the dentist.

This time he examined the right jaw—the pain was on the left side.

"I think I have found out the enemy," said he, as he took the instrument from my mouth with which he had been sounding my teeth. "The corresponding tooth on the other side has commenced decaying, and the nerve is already slightly exposed."

"But what has that to do with this side?" I put my hand where the pain was, as I spoke.

"It may have a good deal to do with it. We shall soon see." And he went to his case of instruments.

"You are not going to extract it, doctor!" I rose from the operating chair in alarm.

"Oh no, no, madam! I am only going to put something into it, to destroy the sensibility of the nerve, previous to preparing it for being filled

The tooth can still be preserved. We will know in a minute or two whether all the difficulty lies here."

A preparation, in which I could perceive the taste and odour of kreosote, was inserted in the cavity of the decayed tooth. In less than five seconds I was free from pain.

"I thought that was it," said the dentist, smiling. "A sound tooth is not very apt to ache of itself. It is sometimes difficult to tell which is the troublesome member. But we have discovered the offending one this time, and will put an end to the disturbance he has been creating."

I could say not a word. My husband looked at me with a humorous expression in his eye. After we were in the street, he remarked, pleasantly—

"No abscess yet, my dear. Were it not for physicians, who understand their business, I am afraid your Job's comforters would soon have you imagine yourself dying, and keep up the illusion until you actually gave up the ghost."

"I really am ashamed of myself," I replied; "but you know how shattered my nerves are, and how little a thing it takes to unsettle me. I do wish my Job's comforters, as you call them, would have more discretion than to talk to me as they do."

"Let them talk; you know it is all talk."

"No—not all talk. They relate real cases of disease and suffering, and I immediately imagine that

I have all the symptoms that ultimately lead to the same sad results."

"Be a woman, Kate! be a woman," responded my husband.

This was all very well, and all easily said. I believe, however, I am a woman, but a woman of the nineteenth century, with nerves far too delicately strung. Ah me! if some of my kind friends would only be a little more thoughtful, they would save me many a wretched day. I hope this will meet the eyes of some of them, and that they will read it to a little profit. It may save others, if it does not save me from a repetition of such things as I have described.

THE CODE OF HONOUR.

Two young men, one with a leather cap on his head and military buttons on his coat, sat in close conversation, long years ago, in the bar-room of the ——— Hotel. The subject that occupied their attention seemed to be a very exciting one, at least to him of the military buttons and black cap, for he emphasized strongly, knit his brow awfully, and at last went so far as to swear a terrible oath.

"Don't permit yourself to get so excited, Tom," interposed a friend. "It won't help the matter at all."

"But I've got no patience."

"Then it is time you had some," coolly returned the friend. "If you intend pushing your way into the good graces of my lady Mary Clinton, you must do something more than fume about the little matter of rivalry that has sprung up."

"Yes; but to think of a poor milk-sop of an author—author?—pah!—scribbler!—to think, I say, of a spiritless creature like Blake thrusting himself between me and such a girl as Mary Clinton; and worse, gaining her notice, is too bad! He has sonneteered her eyebrows, no doubt—flattered her in verse until she don't know who or where she is, and in this way become a formidable rival. But I won't bear it—I'll—I'll"—

"What will you do?"

"Do? I'll—I'll wing him! that's what I'll do. I'll challenge the puppy and shoot him."

And the young lieutenant, for such he was, flourished his right arm and looked pistol-balls and death.

"But he won't fight, Tom."

"Won't he?" and the lieutenant's face brightened. "Then I'll post him for a coward; that'll finish him. All women hate cowards. I'll post him—yes, and cowhide him in the bargain, if necessary."

"Posting will do," half sarcastically replied his friend. "But upon what pretext will you challenge him?"

"I'll make one. I'll insult him the first time I meet him, and then, if he says any thing, challenge and shoot him."

"That would be quite gentlemanly, quite according to the code of honour," returned the friend, quietly.

The young military gentleman we have introduced was named Redmond. The reader has already penetrated his character. In person he was quite good-looking, though not the Adonis he deemed himself. He had fallen deeply in love with the "acres of charms" possessed by a certain Miss Clinton, and was making rapid inroad upon her heart—at least he thought so—when a young man well known in the literary circles made his appearance, and was received with a degree of favour that confounded the officer, who had already begun to think himself sure of the prize. Blake had a much readier tongue and a great deal more in his head than the other, and could therefore, in the matter of mind at least, appear to much better advantage than his rival. He had also written and published one or two popular works; this gave him a standing as an author. Take him all in all, he was a rival to be feared, and Redmond was not long in making the discovery. What was to be done? A military man must not be put down or beaten off by a mere civilian. The rival must be gotten rid of in some manner; the professional means was, as has been seen, thought of first. Blake must be challenged and killed off, and then the course would be clear.

A few days after this brave and honourable determination, the officer met the author in a public place, and purposely jostled him rudely. Blake said nothing, thinking it possible that it was an accident; but he remained near Redmond, to give him a chance to repeat the insult, if such had been his intention. It was not long before the author was again jostled in a still ruder manner than before, at the same time some offensive word was muttered by the officer. This was in the presence of a number of respectable persons, who could not help hearing, seeing, and understanding all. Satisfied that an insult was intended, Blake looked him in the face for a moment, and then asked, loud enough to be heard all around—"Did you intend to jostle me?"

"I did," was the angry retort.

"*Gentlemen* never do such things."

As Blake said this with marked emphasis, he looked steadily into the officer's face.

"You'll hear from me, sir." And as the officer said this, menacingly, he turned and walked away with a military air.

"There's trouble for you now, Blake; he'll challenge you," said two or three friends who instantly gathered around him.

"Do you think so?"

"Certainly; he is an officer—fighting is his trade."

"Well, let him."

"What'll you do?"

"Accept the challenge, of course."

"And fight?"

"Certainly."

"He'll shoot you."

"I'm not afraid."

Blake returned with his friend to his lodgings, where he found a billet already from Redmond, who was all eagerness to wing his rival.

On the next morning, two friends of the belligerents were closeted for the purpose of arranging the preliminaries for the fight.

"The weapon?" asked the friend of the military man. "Your principal, by the laws of honour, has the choice; as, also, to name time and place, &c."

"Yes, I understand. All is settled."

"He will fight, then?"

"Fight? Oh, certainly; Blake is no coward."

"Well, then, name the weapons."

"A pair of goose-quills."

"Sir!" in profound astonishment.

"The weapons are to be a pair of good Russia quills, opaque, manufactured into pens of approved quality. The place of meeting, the —— Gazette; the time, to-morrow morning, bright and early."

"Do you mean to insult me?"

"By no means."

"You cannot be serious."

"Never was more serious in my life. By the code of honour, the challenged party has the right to choose weapons, place of meeting, and time. Is it not so?"

"Certainly."

"Very well. Your principal has challenged mine. All these rights are of course his; and he is justified in choosing those weapons with which he is most familiar. The weapon he can use best is the pen, and he chooses that. If Lieut. Redmond had been the challenged party, he would, of course, have named pistols, with which he is familiar, and Mr. Blake would have been called a coward, poltroon, or something as bad, if, after sending a challenge, he had objected to the weapons. Will your principal find himself in a different position if he decline this meeting on like grounds? I think not. Pens are as good as pistols at any time, and will do as much."

"Fighting with pens! Preposterous!"

"Not quite so preposterous as you may think. Mr. B. has more than insinuated that Mr. Redmond is no gentleman. For this he is challenged to a single combat that is to prove him to be a gentleman or not one. Surely the most sensible weapon with which to do this is the pen. Pistols won't demonstrate the matter; only the pen can do it, so the pen is chosen. In the ——— Gazette of to-morrow morning my friend stands ready to prove that he is a gentleman; and your friend that he is one, and that a gentleman has a right to insult publicly and without provocation whomsoever he pleases. Depend upon it, you will find this quite as serious an affair as if pistols were used."

"I did not come here, sir, to be trifled with."

"No trifling in the matter at all; I am in sober

earnest. Pens are the weapons; the —— Gazette, the battle-ground; time, early as you please to-morrow morning. Are you prepared for the meeting?"

"No."

"Do you understand the consequences?"

"What consequences?"

"Your principal will be posted as a coward before night."

"Are you mad?"

"No, cool and earnest. We fully understand what we are about."

The officer's second was nonplussed; he did not know what to say or think. He was unprepared for such a position of affairs.

"I'll see you in the course of an hour," he at length said, rising.

"Very well; you will find me here."

"Is all settled?" asked the valiant lieutenant, as his second came into his room at the hotel, where he was pacing the floor.

"Settled? No; nor likely to be. I objected to the weapons, and, indeed, the whole proposed arrangement."

"Objected to the weapons! And, pray, what did he name? A blunderbuss?"

"No; nor a duck gun, with trumpet muzzle; but an infernal pen!"

"A what?"

"Why, curse the fellow, a pen! You are to use pens—the place of meeting, the ——— Gazette—time, to-morrow morning. He is to prove you are no gentleman, and you are to prove you are one, and that a gentleman is at all times privileged to insult whomsoever he pleases without provocation."

"He's a cowardly fool!"

"If his terms are not accepted, he threatens tc post you for a coward before night."

"What?"

"You must accept or be posted. Think of that!"

The precise terms in which the principal swore, and the manner in which he fumed for the next five minutes, need not be told. He was called back to more sober feelings by the question—"Do you accept the terms of the meeting?"

"No, of course not; the fellow's a fool."

"Then you consent to be posted. How will that sound?"

"I'll cut off the rascal's ears if he dare do such a thing."

"That won't secure Mary Clinton, the cause of this contest."

"Hang it, no!"

"With pens for weapons he will wing you a little too quick."

"No doubt. But the public won't bear him out in such an outrage—such a violation of all the rules of honour."

"By the code of honour, the challenged party has the right to choose the weapons, &c."

"I know."

"And you are afraid to meet the man you have challenged upon the terms he proposes. That is all plain and simple enough. The world will understand it all."

"But what is to be done?"

"You must fight, apologize, or be posted; there is no alternative. To be posted won't do; the laugh would be too strongly against you."

"It will be as bad, and even worse, to fight as he proposes."

"True. What then?"

"It must be made up somehow or other."

"So I think. Will you write an apology?"

"I don't know; that's too humiliating."

"It's the least of the three evils."

So, at last, thought the valiant Lieut. Redmond. When the seconds again met, it was to arrange a settlement of differences. This could only be done by a very humbly written apology, which was made. On the next day the young officer left the city, a little wiser than he came. Blake and his second said but little about the matter. A few choice friends were let into the secret, which afforded many a hearty laugh. Among these friends was Mary Clinton, who not long after gave her heart and hand to the redoubtable author.

As for the lieutenant, he declares that he had as lief come in contact with a Paixhan gun as an author with his "infernal pen." He understands pistols, small swords, rifles, and even cannons, but he can't stand up when pen-work is the order of the day. The odds would be too much against him.

TREATING A CASE ACTIVELY.

A PHYSICIAN'S STORY.

I WAS once sent for, in great haste, to attend a gentleman of respectability, whose wife, a lady of intelligence and refinement, had discovered him in his room lying senseless upon the floor. On arriving at the house, I found Mrs. H—— in great distress of mind.

"What is the matter with Mr. H——?" I asked, on meeting his lady, who was in tears and looking the picture of distress.

"I'm afraid it is apoplexy," she replied. "I found him lying upon the floor, where he had, to all appearance, fallen suddenly from his chair. His face is purple, and though he breathes, it is with great difficulty."

I went up to see my patient. He had been lifted from the floor, and was now lying upon the bed. Sure enough, his face was purple and his breathing laboured, but somehow the symptoms did not indicate apoplexy. Every vein in his head and face was turgid, and he lay perfectly stupid, but still I saw no clear indications of an actual or approaching congestion of the brain.

"Hadn't he better be bled, doctor?" asked the anxious wife.

"I don't know that it is necessary," I replied. "I think, if we let him alone, it will pass off in the course of a few hours."

"A few hours! He may die in half an hour."

"I don't think the case is so dangerous, madam."

"Apoplexy not dangerous?"

"I hardly think it apoplexy," I replied.

"Pray, what do you think it is, doctor?"

Mrs. H—— looked anxiously into my face.

I delicately hinted that he might, possibly, have been drinking too much brandy; but to this she positively and almost indignantly objected.

"No, doctor; *I* ought to know about that," she said. "Depend upon it, the disease is more deeply seated. I am sure he had better be bled. Won't you bleed him, doctor? A few ounces of blood taken from his arm may give life to the now stagnant circulation of the blood in his veins."

Thus urged, I, after some reflection, ordered a

bowl and bandage, and opening a vein, from which the blood flowed freely, relieved him of about eight ounces of his circulating medium. But he still lay as insensible as before, much to the distress of his poor wife.

"Something else must be done, doctor," she urged, seeing that bleeding had accomplished nothing. "If my husband is not quickly relieved, he must die."

By this time, several friends and relatives, who had been sent for, arrived, and urged upon me the adoption of some more active means for restoring the sick man to consciousness. One proposed mustard plasters all over his body; another a blister on the head; another his immersion in hot water. I suggested that it might be well to use a stomach-pump.

"Why, doctor?" asked one of the friends.

"Perhaps he has taken some drug," I replied

"Impossible, doctor," said the wife. "He has not been from home to-day, and there is no drug of any kind in the house."

"No brandy?" I ventured this suggestion again.

"No, doctor, no spirits of any kind, nor even wine, in the house," returned Mrs. H——, in an offended tone.

I was not the regular family physician, and had been called in to meet the alarming emergency, because my office happened to be nearest to the dwelling of Mr. H——. Feeling my position to be a difficult one, I suggested that the family physician had better be called.

"But the delay, doctor," urged the friends.

"No harm will result from it, be assured," I replied.

But my words did not assure them. However, as I was firm in my resolution not to do any thing more for the patient until Dr. S—— came, they had to submit. I wished to make a call of importance in the neighbourhood, and proposed going, to be back by the time Dr. S—— arrived; but the friends of the sick man would not suffer me to leave the room.

When Dr. S—— came, we conversed aside for a few minutes, and I gave him my views of the case, and stated what I had done and why I had done it. We then proceeded to the bedside of our patient; there were still no signs of approaching consciousness.

"Don't you think his head ought to be shaved and blistered?" asked the wife, anxiously.

Dr. S—— thought a moment, and then said—"Yes, by all means. Send for a barber; and also for a fresh fly-blister, four inches by nine."

I looked into the face of Dr. S—— with surprise; it was perfectly grave and earnest. I hinted to him my doubt of the good that mode of treatment would do; but he spoke confidently of the result, and said that it would not only cure the disease, but, he believed, take away the predisposition thereto, with which Mr. H—— was affected in a high degree.

The barber came. The head of H—— was shaved, and Dr. S—— applied the blister with his own hands, which completely covered the scalp from forehead to occiput.

"Let it remain on for two hours, and then make use of the ordinary dressing," said Dr. S——. "If he should not recover during the action of the blister, don't feel uneasy; sensibility will be restored soon after."

I did not call again, but I heard from Dr. S—— the result.

After we left, the friends stood anxiously around the bed upon which the sick man lay; but though the blister began to draw, no signs of returning consciousness showed themselves, further than an occasional low moan, or an uneasy tossing of the arms. For full two hours the burning plaster parched the tender skin of H——'s shorn head, and was then removed; it had done good service. Dressings were then applied; repeated and repeated again; but still the sick man lay in a deep stupor.

"It has done no good; hadn't we better send foı the doctor?" suggested the wife.

Just then the eyes of H—— opened, and he looked with half-stupid surprise from face to face of the anxious group that surrounded the bed.

"What in the mischief's the matter?" he at length said. At the same time, feeling a strange sensation about his head, he placed his hand rather heavily thereon.

"Heavens and earth!" He was now fully in his senses. "Heavens and earth! what ails my head?"

"For mercy's sake, keep quiet," said the wife the glad tears gushing over her face. "You have been very ill; there, there, now!" And she spoke soothingly. "Don't say a word, but lie very still."

"But my head! What's the matter with my head? It feels as if scalded. Where's my hair? Heavens and earth! Sarah, I don't understand this. And my arm? What's my arm tied up in this way for?"

"Be quiet, my dear husband, and I'll explain it all. Oh, be very quiet; your life depends upon it."

Mr. H—— sank back upon the pillow from which he had arisen, and closed his eyes to think. He put his hand to his head, and felt it, tenderly, all over, from temple to temple, and from nape to forehead.

"Is it a blister?" he at length asked.

"Yes, dear. You have been very ill; we feared for your life," said Mrs. H——, affectionately; "there have been two physicians in attendance."

H—— closed his eyes again; his lips moved Those nearest were not much edified by the whis-

pered words that issued therefrom. They would have sounded very strangely in a church, or to ears polite and refined. After this, he lay for some time quiet.

"Threatened with apoplexy, I suppose?" he then said, interrogatively.

"Yes, dear," replied his wife. "I found you lying insensible upon the floor, on happening to come into your room. It was most providential that I discovered you when I did, or you would certainly have died."

H—— shut his eyes and muttered something, with an air of impatience; but its meaning was not understood. Finding him out of danger, friends and relatives retired, and the sick man was left alone with his family.

"Sarah," he said, "why, in the name of goodness, did you permit the doctors to butcher me in this way? I'm laid up for a week or two, and all for nothing."

"It was to save your life, dear."

"Save the ——!"

"H-u-s-h! There! do, for mercy's sake, be quiet; every thing depends upon it."

With a gesture of impatience, H—— shut his eyes, teeth, and hands, and lay perfectly still for some minutes. Then he turned his face to the wall, muttering in a low, petulant voice—"Too bad! too bad! too bad!"

I had not erred in my first and my last impressions of H——'s disease, neither had Dr. S——, although he used a very extraordinary mode of treatment. The facts of the case were these :

H—— had a weakness; he could not taste wine nor strong drink without being tempted into excess. Both himself and friends were mortified and grieved at this; and they, by admonition, and he, by good resolutions, tried to bring about a reform; but to see was to taste, to taste was to fall. At last, his friends urged him to shut himself up at home for a certain time, and see if total abstinence would not give him strength. He got on pretty well for a few days, particularly so, as his coachman kept a well-filled bottle for him in the carriage-house, to which he not unfrequently resorted; but a too ardent devotion to this bottle brought on the supposed apoplexy.

Dr. S—— was right in his mode of treating the disease after all, and did not err in supposing that it would reach the predisposition. The cure was effectual. H—— kept quiet on the subject, and bore his shaved head upon his shoulders with as much philosophy as he could muster. A wig, after the sores made by the blister had disappeared, concealed the barber's work until his own hair grew again. He never ventured upon wine or brandy again for fear of apoplexy.

When the truth leaked out, as leak out such

things always will, the friends of H—— had many a hearty laugh; but they wisely concealed from the object of their merriment the fact that they knew any thing more than appeared of the cause of his supposed illness.

THE END.

By Miss MARGARET LEE.

*Divorce.........................20

By HENRY W. LONGFELLOW.

*Hyperion.........................20
*Outre-Mer.........................20

By SAMUEL LOVER.

The Happy Man.........................10

By LORD LYTTON.

The Coming Race.........................10
Leila, or the Siege of Granada.........................10
Earnest Maltravers.........................20
The Haunted House, and Calderon the Courtier.........................10
Alice; a sequel to Earnest Maltravers.20
A Strange Story.........................20
*Last Days of Pompeii.........................20
Zanoni.........................20
Night and Morning, Part I.........................15
" " Part II.........................15
Paul Clifford.........................20
Lady of Lyons.........................10
Money.........................10
Richelieu.........................10

By H. C. LUKENS,

*Jets and Flashes.........................20

By Mrs. E. LYNN LINTON.

Ione Stewart.........................20

By W. E. MAYO.

The Berber.........................20

By A. MATHEY.

Duke of Kandos.........................20
The Two Duchesses.........................20

By JUSTIN H. McCARTHY.

An Outline of Irish History.........................10

By EDWARD MOTT.

*Pike County Folks.........................20

By MAX MULLER.

*India, what can she teach us?.........................20

By Miss MULOCK.

*John Halifax.........................20

By R. HEBER NEWTON

The Right and Wrong Uses of the Bible.........................20

By W. E. NORRIS.

*No New Thing.........................20

By OUIDA.

*Wanda, Part I.........................15
" Part II.........................15
*Under Two Flags, Part I.........................20
" " Part II.........................20

By Mrs. OLIPHANT.

*The Ladies Lindores.........................20

By LOUISA PARR.

Robin.........................20

By JAMES PAYN.

*Thicker than Water.........................20

By CHARLES READE.

Single Heart and Double Face.........................10

By REBECCA FERGUS REDCLIFF.

Freckles.........................20

By Sir RANDALL H. ROBERTS.

Harry Holbrooke.........................20

By Mrs. ROWSON.

Charlotte Temple.........................10

By W. CLARK RUSSELL.

*A Sea Queen.........................20

By GEORGE SAND.

The Tower of Percemont.........................20

By Mrs. W. A. SAVILLE.

Social Etiquette.........................15

By MICHAEL SCOTT.

*Tom Cringle's Log.........................20

By EUGENE SCRIBE.

Fleurette.........................20

By J. PALGRAVE SIMPSON.

Haunted Hearts.........................10

By GOLDWIN SMITH, D.C.L.

False Hopes.........................15

By DEAN SWIFT

Gulliver's Travels.........................20

By W. M. THACKERAY.

*Vanity Fair, Part I.........................15
" " " II.........................15

By Judge D. P. THOMPSON.

*The Green Mountain Boys.........................20

By THEODORE TILTON.

Tempest Tossed, Part I.........................20
" " Part II.........................20

By JULES VERNE.

*800 Leagues on the Amazon.........................10
*The Cryptogram.........................1[illegible]

By GEORGE WALKER.

*The Three Spaniards.........................20

By W. M. WILLIAMS.

Science in Short Chapters.........................20

By Mrs. HENRY WOOD.

*East Lynne.........................20

MISCELLANEOUS.

Paul and Virginia.........................10
Margaret and her Bridesmaids.........................20
The Queen of the County.........................20
Baron Munchausen.........................10

www.ingramcontent.com/pod-product-compliance
Lightning Source LLC
LaVergne TN
LVHW050530100826
845148LV00002B/505